10

Band 22 · Oktober 1986

isotopenpraxis

AKADEMIE-VERLAG BERLIN

ISSN 0021-1915 Isotopenpraxis, Berlin 22 (1986) 10, 337–376

An unsere Autoren

Die Zeitschrift veröffentlicht Beiträge zu folgenden Themen: Kontrolle und Analyse radioaktiver und stabiler Isotope, Anwendung radioaktiver und stabiler Isotope in allen Gebieten der Forschung und Technik, der Industrie, Landwirtschaft und Medizin, Herstellung radioaktiver und stabiler Isotope sowie von markierten Verbindungen, kernelektronische Geräte für die Isotopenanwendung (Dickenmesser, Dichtemesser u. ä.) Detektoren für isotopentechnische Arbeiten und den Strahlenschutz in Isotopenlaboratorien, Meßtechnik und Strahlenschutz, Dosimetrie in Isotopenlaboratorien, kerntechnische Geräte für die Isotopenanwendung (Beton, Bestrahlungsanlagen u. ä.), Strahlenchemie, Radiochemie, ökonomische und perspektivische Fragen, Fragen der Weiterbildung.

1. Aufnahme von Manuskripten

Zur Veröffentlichung angenommen werden nur druckreife, noch nicht anderweitig publizierte Beiträge in deutscher, englischer oder russischer Sprache zum Themenbereich der Zeitschrift und zwar

Originalarbeiten
Kurze Mitteilungen
Aus der Praxis
Zusammenfassende Berichte
Tagungsberichte
Informationen.

Sind der Beitrag oder Teile daraus bereits veröffentlicht oder ist eine anderweitige Veröffentlichung vorgesehen, so muß dies der Redaktion mit genauer Angabe der Zweitveröffentlichung mitgeteilt werden. Mit der Annahme des Manuskriptes behält sich der Verlag das Recht der Vervielfältigung, Verbreitung und Übersetzung innerhalb der gesetzlichen Schutzfrist vor. Auszugsweiser Nachdruck oder photomechanische u. a. Wiedergabe sind an eine schriftliche Genehmigung des Verlages gebunden. Manuskripte, Zeichnungen und Photos von veröffentlichten Arbeiten werden nur auf den besonderen und bis zum Zeitpunkt der Veröffentlichung mitgeteilten Wunsch des Autors zurückgesandt.
Bei der unverlangten Einreichung von Manuskripten wird empfohlen, eine Befürwortung eines der Herausgeber oder Redaktionsausschußmitglieder beizufügen.
Manuskripte sind einzureichen bei der Redaktion, 7010 Leipzig, PSF 370.

2. Form der Manuskripte

Die Autoren werden gebeten:
2 Manuskripte, eines davon in Original-Maschinenschrift, einseitig und mit 2zeiligem Abstand einzureichen; gut leserliche Photokopien sind zulässig; auf dem 1. Blatt (unter Weglassung der Bezeichnung)
den Titel der Arbeit
Vorname-Initialen und Namen sämtlicher Autoren
gegebenenfalls die Institution, in der sie tätig sind
eine kurze Zusammenfassung der Arbeit von höchstens 15 Zeilen
Keywords in alphabetischer Reihenfolge und
als Fußnote die Anschrift der Institution anzugeben;
wenn möglich, Zusammenfassung, Titel und Autoren *englisch* (bei deutschem oder russischen Manuskript) bzw. *deutsch* (bei englischem Manuskript) 2fach beizufügen;
Literaturhinweise (fortlaufend numeriert), Fußnoten, Tabellen, Abbildungen (in 2–3facher Veröffentlichungsgröße), Abbildungsunterschriften (getrennt von den Abbildungen und Zeichenerklärungen, Quellenangaben, falls erforderlich, und gegebenenfalls mit einem Vermerk für besondere Wünsche bezüglich der Abbildungsgröße), private und Dienstanschrift der Verfasser mit akademischen Titeln und ausgeschriebenen Vornamen auf besonderen Blättern ans Ende des Manuskriptes zu stellen;
Wünsche für den Kolumnentitel (Seitenüberschrift) bekanntzugeben;
Absätze durch Einzüge von mindestens 3 Anschlägen kenntlich zu machen;
die Maßeinheiten des SI-Systems („Systeme International") zu verwenden;
Korrekturen im Manuskript unmittelbar an der zu korrigierenden Stelle anzubringen;
Formeln deutlich lesbar und möglichst groß einzutragen;
Farbabbildungen und nachträgliche Änderungen oder Ergänzungen in Abbildungen zu vermeiden;
hervorzuhebende Textstellen zu unterstreichen, und zwar halbfetten Satz mit einer geraden schwarzen Linie, kursiven Satz (Schrägschrift) mit einer blauen Wellenlinie (alle Eigennamen). Sperrungen gestrichelt;
besondere Buchstaben durch farbige Unterstreichung zu kennzeichnen, und zwar Formelzeichen in

Griechisch-Kursiv (α, β, ...) rot,
Kursiv (*A*, *B*, ...) blau
halbfetter Kursiv (**A**, **B**, ...) 2 Linien blau (Vektoren),
halbfetter Grotesk (**A**, **B**, ...) 2 Linien violett (Tensoren);

bei Literaturangaben auf Vollständigkeit zu achten. In nachstehender Reihenfolge müssen angegeben sein:

a) bei *Zeitschriften*: Vorname-Initialen und Namen sämtlicher Autoren, Titel der Zeitschrift (nach „Periodica Chimica" oder möglichst wenig gekürzt), Bandnummer (unterstrichen), Jahreszahl (in Klammern), Heftnummer (nur bei Zeitschriften, die jedes Heft mit S. 1 beginnen) sowie die Seitenzahl, auf der die zitierte Arbeit beginnt.

b) bei *Büchern*: Initialen der Vornamen und Name des Autors (Herausgeber werden durch den Zusatz „ed." gekennzeichnet), voll ausgeschriebener Titel des Buches in der Originalsprache, evtl. Bandnummer, voll ausgeschriebener Name des Verlages, Nummer der Auflage (außer der ersten), Erscheinungsort(e), Erscheinungsjahr (hier nicht in Klammern setzen!) und evtl. Kapitelnummer, Seitenzahl u. ä.

c) bei *Sammelbänden, Tagungsberichten*: Es soll zusätzlich der Titel des zitierten Beitrags angegeben werden. Die Seitenzahl muß unbedingt angegeben werden. Aus dem Zitat muß hervorgehen, ob es sich um einen Tagungssammelband oder um einen noch nicht veröffentlichten Vortrag handelt. Zwischen Tagungsort und Erscheinungsjahr des entsprechenden Tagungsberichts ist zu unterscheiden.

3. Korrekturen

Der Autor erhält zwei Umbruchsabzüge, von denen einer möglichst schnell korrigiert an die Redaktion zurückgereicht werden soll. Auf Wunsch des Autors kann die Korrektur auch von der Redaktion gelesen werden.

Durch nachträgliche Manuskriptänderungen verursachte Korrekturkosten gehen, sofern sie 10% der Satzkosten übersteigen, zu Lasten des Autors.

Der Autor erhält 50 Sonderdrucke kostenlos. Weitere Sonderdrucke können bis zur Rückgabe der Korrekturabzüge auf Kosten des Autors bei der Redaktion bestellt werden.

Bezugsmöglichkeiten

Bestellungen sind zu richten

— **in der DDR** an den Postzeitungsvertrieb unter Angabe der Kundennummer des Bestellenden oder an den AKADEMIE-VERLAG BERLIN, DDR-1086 Berlin, Leipziger Straße 3–4, PF-Nr. 1233;
— **im sozialistischen Ausland** an eine Buchhandlung für fremdsprachige Literatur oder an den zuständigen Postzeitungsvertrieb;
— **in der BRD und Berlin (West)** an eine Buchhandlung oder an die Auslieferungsstelle KUNST UND WISSEN, Erich Bieber OHG, Wilhelmstraße 4–6, D-7000 Stuttgart 1;
— **in den übrigen westeuropäischen Ländern** an eine Buchhandlung oder an die Auslieferungsstelle KUNST UND WISSEN, Erich Bieber GmbH, General Wille-Str. 4, CH-8002 Zürich;
— **im übrigen Ausland** an den Internationalen Buch- und Zeitschriftenhandel; den Buchexport Volkseigener Außenhandelsbetrieb der Deutschen Demokratischen Republik, DDR-7010 Leipzig, Postfach 160; oder an den AKADEMIE-VERLAG BERLIN, DDR-1086 Berlin, Leipziger Straße 3–4.

Zeitschrift „Isotopenpraxis"
Herausgeber: Zentralinstitut für Isotopen- und Strahlenforschung der Akademie der Wissenschaften der DDR.
Verlag: Akademie-Verlag Berlin, DDR-1086 Berlin, Leipziger Straße 3–4, PF-Nr. 1233; Fernruf: 223621 oder 2236229; Telex-Nr.: 114420;
Bank: Staatsbank der DDR, Berlin, Kto.-Nr.: 6836-26-20712.
Chefredakteur: Prof. Dr. H. Hübner, Leipzig.
Anschrift der Redaktion der Zeitschrift Isotopenpraxis: Prof. Dr. H. Hübner, DDR-7010 Leipzig, Karl-Marx-Platz, Hauptpostamt, PSF 370; Fernruf: 281018.
Veröffentlicht unter der Lizenznummer 1527 des Presseamtes beim Vorsitzenden des Ministerrates der Deutschen Demokratischen Republik.
Gesamtherstellung: VEB Druckerei „Thomas Müntzer", DDR-5820 Bad Langensalza.
Erscheinungsweise: Die Zeitschrift „Isotopenpraxis" erscheint monatlich. Die 12 Hefte einer Jahrganges bilden einen Band. Bezugspreis je Band 240,— DM zuzüglich Versandspesen; Preis je Heft 20,— DM. Der gültige Jahresbezugspreis für die DDR ist der Postzeitungsliste zu entnehmen.
Bestellnummer dieses Heftes: 1079/22/10.

AN (EDV) 12831
00900

Isotopenpraxis 22 (1986) 10, S. 337 bis 344

Radionuclides in Environmental Research and Protection*)

W. Michaelis

(Institute of Physics, GKSS Research Centre Geesthacht, F.R.G.)[1]

Radioisotopes of both natural and artificial origin are widely used in environmental research and protection to tackle problems in which high specificity and sensitivity are required. The paper exemplifies the capabilities guided by some recent investigations in various fields of application: dating of river and marine sediments, determination of dry deposition rates for airborne particulates, studies of the uptake of organic compounds by plants and of the internal distribution processes in plants, investigation of transport phenomena in soil, groundwater or surface waters, development of trace analytical procedures, application of neutron activation analysis, detection of trace gases by means of ionization detectors, and the exploration of mud fields in tidal rivers using a radiometric density gauge.

Natürliche und künstliche Radioisotope werden vielseitig in der Umweltforschung und im Umweltschutz angewendet, wo eine hohe Spezifität und Empfindlichkeit zur Lösung von Problemen benötigt werden. Der Beitrag illustriert an einigen Beispielen die Möglichkeiten der Anwendungen, wie sie sich in letzter Zeit aus Untersuchungen auf unterschiedlichen Gebieten ergeben haben: Datierung von Fluß- und Meeressedimenten, Bestimmung der trockenen Depositionsraten von festen Partikeln aus der Atmosphäre, Untersuchungen über die Aufnahme organischer Verbindungen durch Pflanzen und über interne Verteilungsvorgänge in Pflanzen, Untersuchungen über Transportvorgänge in der Bodenzone, im Oberflächen- und Grundwasser, Entwicklung von spurenanalytischen Verfahren, Anwendung der Neutronenaktivierungsanalyse, Bestimmung von Spurengasen mit Hilfe von Ionisationsdetektoren und die Erforschung von Schlickgebieten im Gezeitenbereich von Flüssen unter Verwendung radiometrischer Dichtemessungen.

Keywords

age estimation; air pollution; environment; neutron activation analysis; organic chlorine compounds; plants; radiometric gages; radionuclide kinetics; rivers; sediments; tracer techniques; uptake

1. Introduction

The conservation of the environment is one of the most important present-day problems. For a long time, during the rapid development of industry and the enormous productivity intensification in agriculture, economic aspects have been nearly exclusively in the foreground. During the last decade, however, it has been realized in many countries that equal significance is due to ecological points of view if grave impacts on vegetation, animals and human beings are to be prevented.

The tasks which have to be solved in environmental research and protection are found to be of manifold nature. Radionuclides often provide a powerful means to tackle problems in which high specificity and sensitivity are required. These properties of radioisotopes are due to the facts that materials, molecules or elements are or can be labelled in a definite or well-aimed manner, that the emitted radiation is very specific, and that radioactivity can be detected and tracked with high sensitivity. Other applications are based on ionization effects or take advantage of the interaction and attenuation of nuclear radiation.

The present paper exemplifies the capabilities of the use of radioisotopes in environmental research and protection by some recent applications in quite different fields.

*) Plenary lecture presented at the 3rd Working Meeting on "Radioisotope Application and Radiation Processing in Industry" held in Leipzig, G.D.R., September 23 – 29, 1985

[1]) Address: Max-Planck-Str., D-2054 Geesthacht, P.O. Box 1160

2. Dating of sediments

The assessment of the degree of anthropogenic pollution of surface waters is facilitated if the pollution history and the geochemical background are known. This information can be obtained by coupling trace analysis and dating of sediment depth profiles. Investigations of this kind have recently been performed by *Anders* et al. [1], *Spychala* [2], and *Schönburg* et al. [3] on the Elbe River. Sediment cores were taken with a gravity corer which was operated from board a ship. Heavy metal trace analysis of the strata was performed using total-reflection X-ray fluorescence (TXRF) [4–6], prompt gamma neutron activation analysis (PGNAA) [2, 7], and inductively coupled plasma optical emission spectroscopy (ICP) [8], respectively.

Pollution chronology may be derived from low-level counting of ^{137}Cs released in nuclear weapon tests or of ^{210}Pb which is a member of the uranium decay series [9]. The atmospheric flux of ^{137}Cs increased during the early fifties and reached pronounced maxima in 1959 and 1963 [10]. An example for the depth dependence of some heavy metal concentrations in the < 63 µm grain-size fraction, of the percentage of grain-size < 63 µm, and of the ^{137}C, activity in a sediment core from the Elbe River is shown in Fig. 1 [1]. The region of increased concentrations of Cu, Zn, As, Cd, and Pb extends approximately over twice the depth exhibiting appreciable ^{137}Cs activities. This observation suggests that pollution started long before the numerous nuclear weapon tests in the atmosphere during the

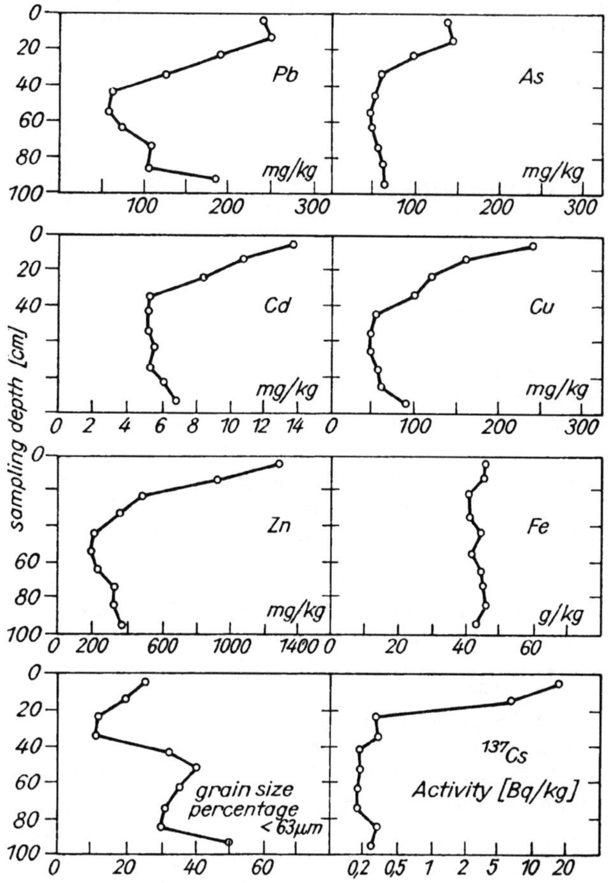

Fig. 1. Element concentrations in a sediment core from the Elbe River

fifties which is in good agreement with studies of *Dominik* et al. [11] on marine sediments from the German Bight (North Sea). Unfortunately, in large rivers and estuaries the formation of the strata is not equally undisturbed as in the marine environment or in lakes since bioturbation, extreme upper water, heavy tidal waves or dredging work may destroy the regular stratification from the sedimentation rates.

The ^{137}Cs activity shown in Fig. 1 for sampling depths between 40 and 80 cm corresponds to the background radiation which for the apparatus used is equivalent to a little less than 0·2 Bq/kg. The slight increase of both heavy metal concentrations and ^{137}Cs activity below 80 cm is ascribed to a porous layer which allowed infiltration of polluted material into the deeper strata.

The natural radioisotope ^{210}Pb is deposited with settling suspended particulate matter and decays in the sediment with its half-life of 22·3 a. The use of this nuclide, therefore, allows dating of sediments from the beginning of the industrial era. In earlier investigations rather complex separation techniques were applied to detect ^{210}Pb or its daughter ^{210}Po by beta or alpha counting (see, e.g., [12]). Lately, *Larsen* and coauthors [13] have used nondestructive low-energy photon counting to quantify ^{210}Pb concentrations in sediments via the 46 keV gamma radiation. They employed an intrinsic Ge planar detector having a thin beryllium entrance window. Good agreement with conventional chemical techniques was found in the range 0 to 1·8 Bq/g. Similar attempts are undertaken at the author's institute by means of an intrinsic Ge borehole detector with a thin magnesium cap in a heavily shielded low-level apparatus. A typical gamma-ray spectrum of a sediment sample is presented in Fig. 2.

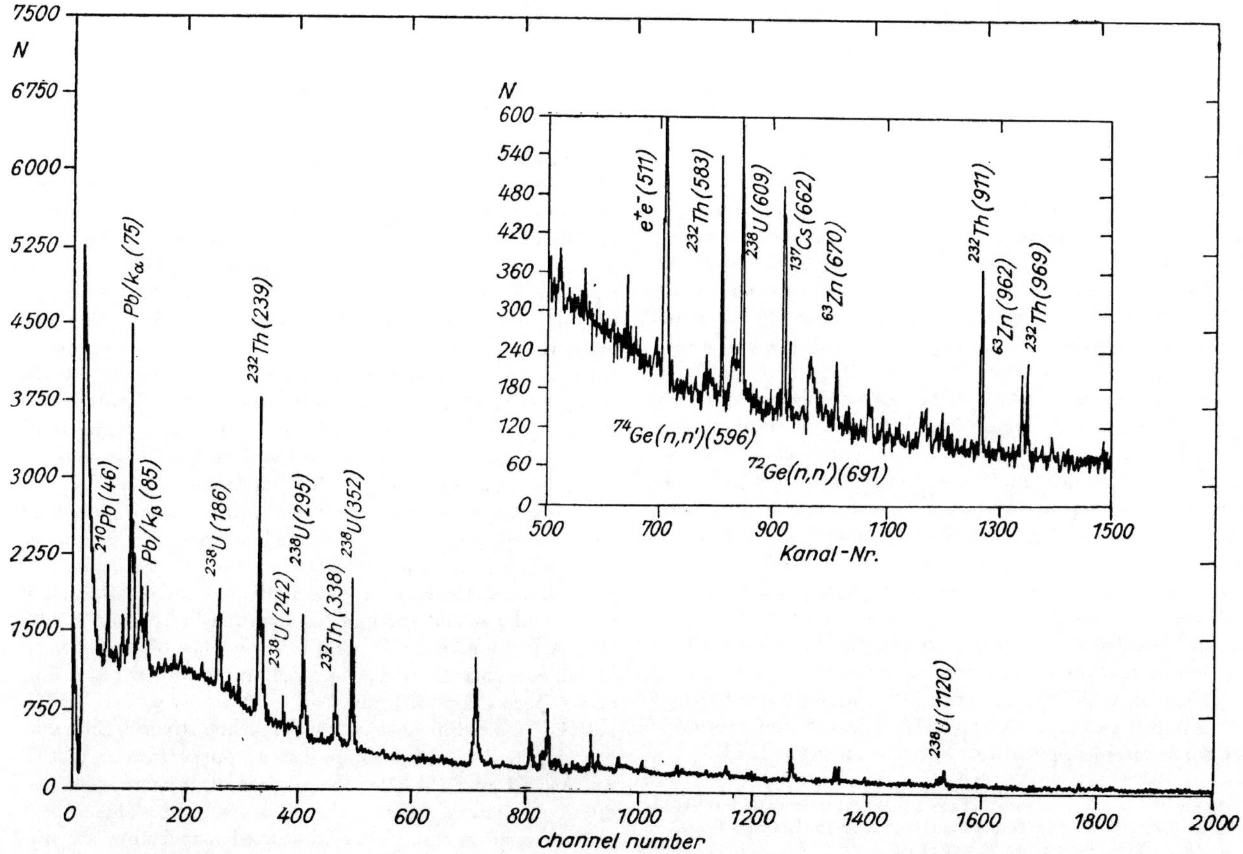

Fig. 2. Gamma-ray spectrum of a sediment sample. Sampling depth 55 to 28 cm below the water/sediment interface

3. Deposition velocities of airborne particulates

Heavy metals are deposited onto soil and vegetation by rainwater, snow, mist, dew, rime and particulates. The recognition of the potential impact on the environment and of the gap of knowledge on the processes involved is presently stimulating increased efforts in the research on wet and dry deposition phenomena [14—18].

One of the most promising methods for the determination of dry deposition rates of airborne particulates and heavy metals bound to them is the so-called 'concentration method' in which size-fractionated aerosol samples are collected by means of a high-volume sampler equipped with a multistage impactor [14]. The loaded substrates are then subjected to trace element analysis, using, for instance, total-reflection X-ray fluorescence analysis (TXRF) [7, 14] or atomic absorption spectrometry (AAS) [19]. The flux F to the surface can be determined by means of the equation

$$F = \sum_i c_i v_i \,.$$

Here, c_i and v_i are the particle-size dependent heavy metal concentration in air and the deposition velocity for particle-size i, respectively. The deposition velocities vary in the range 0·5 to 10 µm by more than two orders of magnitude. They depend on particle variables, but also on surface properties, roughness heights and friction velocities.

Jonas and *Vogt* [16] and *Jonas* [17] have used radioactivity-labelled aerosols to measure deposition velocities in extensive field experiments. Various interfaces have been considered: soil without vegetation, water, filter paper, smooth and structured metals, grass, clover and trees. Polydisperse and monodisperse test aerosols (dia. from 0·4 µm to 17 µm) were produced by a two-fluid nozzle generator and a Berglund-Liu generator. Reactor-activated copper sulphate was used as a tracer.

Other authors [20] applied pyrotechnically produced aerosols, with rare earth elements as tracer, to study the variation of particle-size distribution during atmospheric dispersion processes. The particulates were collected by impactors at different distances from the source. The tracer can easily be detected with high sensitivity by neutron activation.

4. Uptake of agents by plants

Plant-protective agents provide an important means in agriculture to intensify the productivity. However, the fate of the various chemicals that are employed must be known in order to avoid implications injurious to health within the food chain.

This involves, in particular, uptake by plants, distribution processes in the plants, accumulation, excretion, metabolization, and formation of unextractable residues. The research in this field depends to a high degree upon the application of ^{14}C-labelled compounds and the sensitive detection of this radioisotope.

Führ [21] and *Führ* and *Steffens* [22] have studied the uptake and translocation of ^{14}C-labelled fungicides. Detailed information on uptake pathways and distribution processes in the plants were obtained by macroautoradiographing whole plants and microautoradiographing tissue sections [23, 24]. One of the important results of this research is that indications of selective applications of the active substances can be elaborated with the aim to reduce the volumes of plant-protective agents.

The exploration of the fate of various widely distributed organic pollutants and potential substitutes for industrial chemicals in soil-plant systems has been the objective of numerous recent investigations. These studies, in particular, include polychlorinated biphenyls (PCBs) [25, 26], chloroalkylenes [27], pentachlorophenol [28], aldrin and its metabolite dieldrin [28, 29], and chloroanilines which may be formed from herbicides in the environment [30, 31].

Besides ^{14}C, the radionuclide ^{11}C is of practical importance in studies on biological systems since it allows in vivo measurements by means of the 511 keV positron annihilation radiation. The isotope can be produced with high specific activity in cyclotrons via the $^{14}N(p, \alpha)^{11}C$ reaction in a nitrogen gas target. The half-life (20·3 min) is rather short, but it permits repeated measurements at the same system. *Jahnke* et al. [32] and *Wieneke* et al. [33] have successfully used $^{11}CO_2$ to investigate the assimilate transport in plants. The latter authors also proposed applications of the method to plant-physiological studies on environmental impacts of increased concentrations of gases like CO_2, SO_2, O_3, NO_x or of noxious airborne particulates.

5. Tracer techniques and pollutant transport

Transport phenomena in soil, groundwater or surface waters may effectively be explored by using radioactive isotopes or stable tracers which are activated after sampling and are then determined with high sensitivity. Applications in hydrology have been reviewed a few years ago by *Moser* and *Rauert* [34] and by the IAEA/UNESCO Working Group on Nuclear Techniques in Hydrology [35].

The groundwater quality in the sphere of influence of industrial or agricultural sources of pollutants crucially depends on the hydrogeological conditions. The protection of groundwater and the selection of sites for potential pollutant emitters, therefore, requires a detailed knowledge of filtration velocity and flow direction. Borehole techniques for determining these aquifer characteristics have, e.g., been developed by *Drost* and *Klotz* [36]. They are based on dilution logging after tracer injection and the rotation of the detector collimator. The measured velocities and directions may be used to calculate the transmissivity and the dispersivity of the aquifer which are the most important parameters for assessing the pollutant transport.

Groundwater tracers have to fulfill several requirements. They must be representative of the groundwater, they must not undergo chemical reactions with water and aquifer, and their introduction must not disturb the flow pattern (*Drost* and *Klotz* [36]). These criteria are met by the gamma-ray emitting isotopes ^{82}Br (as NH_4Br solution), ^{51}Cr (as EDTA complex), ^{58}Co and ^{60}Co (as $Co(CN)_6$ complex), ^{131}I (as NaI solution), and ^{99}Tc (pertechnate as eluate from ^{99m}Tc generator). Tritiated water is, in principle, the most ideal hydrological tracer, but a drawback is that it emits β-radiation only. Radioactive logging methods such as gamma logging (GL), gamma-gamma logging (GGL), neutron-neutron logging (NNL), and neutron-gamma logging (NGL) may be used to obtain qualitative estimates of the rock composition, the permeability, the bulk density, and the hydrogen content (*Ferronsky* and *Czubek* [37]).

One of the most sensitive activatable tracers is iridium since this element can be determined with excellent detection power by neutron activation analysis and since its natural level is extremely low. The dilution factor, therefore, may be in the order of 10^{12}. *Schnier* and *Fanger* [38] have reported on the application of Ir-labelled tailings from the floatation process aboard a ship during a prepilot mining test for hydrothermal slimes in the Red Sea. The tailings were disposed at a water depth of 400 m. The objective of the experiment was to track the low-grade, but still noxious material by taking, after settling, samples from the sea-bed using box corers and free-fall grabs. The dispersion of the tailings should be known in order to exclude ecological hazards for the coastal regions populated by coral and fish. The sediment samples were analysed by radiochemical neutron activation analysis. The detection limits for iridium were in the range (2.5 to 5) $\times 10^{-12}$ g per g dry substance.

6. Development of trace analytical procedures

In environmental research and protection trace elements have to be determined over a wide range of atomic number, down to very low concentrations, and in quite different matrices. This challenge requires the availability of complementary analytical methods which are characterized by a high detection power and minor sources of systematic

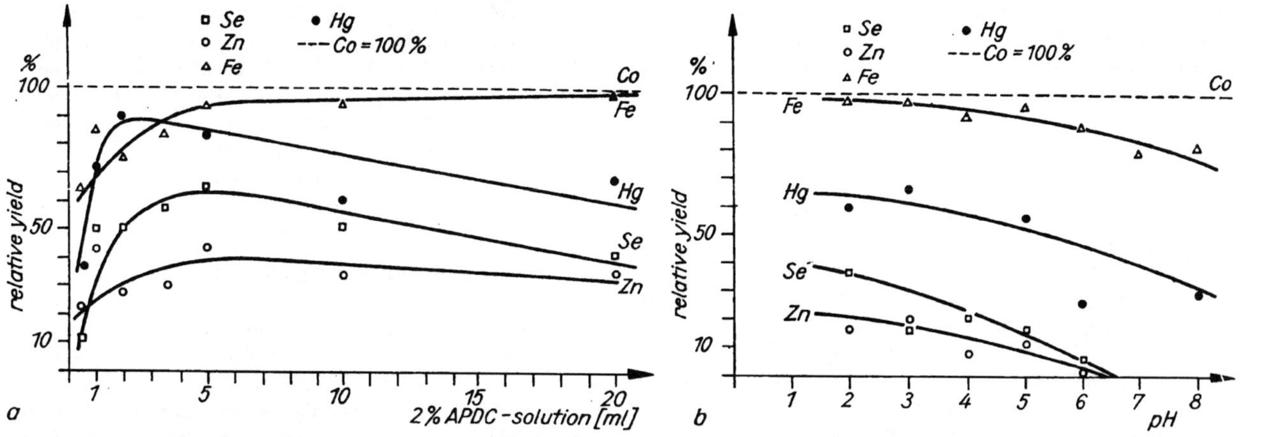

Fig. 3. Tracer experiment with APDC as a chelating ligand. Recovery relative to Co; ph = 5 in the left figure

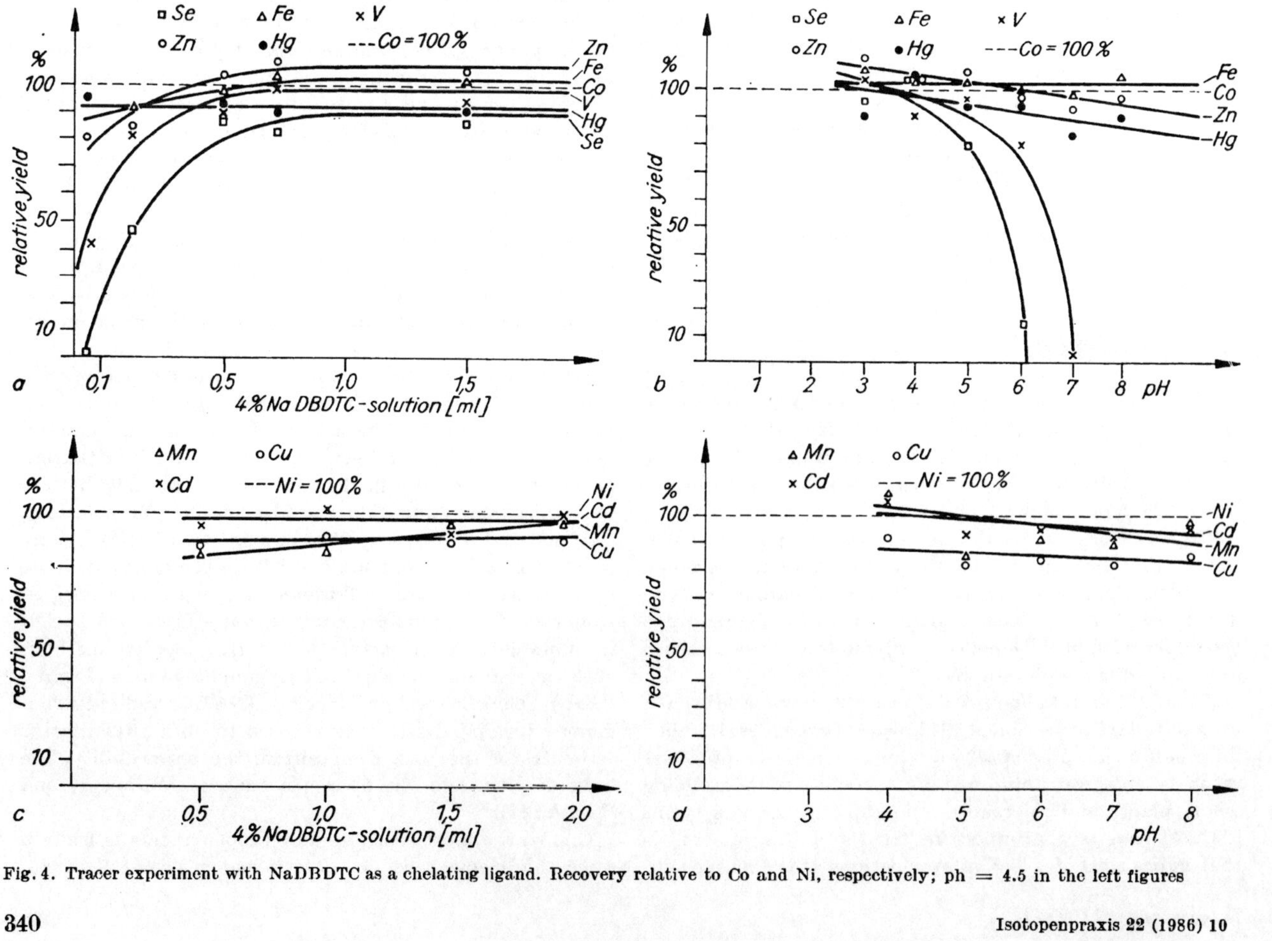

Fig. 4. Tracer experiment with NaDBDTC as a chelating ligand. Recovery relative to Co and Ni, respectively; ph = 4.5 in the left figures

errors. Often only coupling several methods in a concerted attack represents a promising way to cope with a given analytical problem. Radionuclides can effectively be utilized in the development of new procedures. This shall be exemplified by means of the examination and optimization of a novel matrix separation and preconcentration technique for total-reflection X-ray fluorescence analysis (*Prange* [39], *Prange* et al. [40]).

Though TXRF is a rather new technique, its use is increasing rapidly because of the low sample masses required, the simple calibration procedures, and the high detection power [4—7].

For several matrices best performance is achieved by applying appropriate matrix separation and trace element preconcentration. Chelation with dithiocarbamates, selective chromatographic adsorption of the metal complexes, and subsequent elution of the chelates by organic solvents is one of the sample preparation techniques used. The method described by *Prange* was originally developed for sea-water, but later on it was also modified for the application to rainwater [14, 41].

Since complexation of heavy metals with dithiocarbamates is known to be sensitive to pH and concentration, the recovery of the traces, as a function of these parameters and the type of complexing agent employed, requires careful consideration. Radionuclides such as ^{48}V, ^{51}Cr, ^{54}Mn, ^{56}Mn, ^{59}Fe, ^{60}Co, ^{65}Ni, ^{64}Cu, ^{65}Zn, ^{74}As, ^{75}Se, ^{115}Cd, and ^{203}Hg may be used for these studies. In Figs. 3 and 4 the results of tracer experiments with ammonium pyrrolidinedithiocarbamate (APDC) and sodium dibenzyldithiocarbamate (NaDBDTC) as chelating ligands are shown. It is demonstrated that the latter agent leads to clearly better yields, that the NaDBDTC concentration should be higher than 14 μmol/kg, and that the pH value must not exceed 5.

7. Neutron activation analysis

A survey on radionuclides in environmental research and protection would be incomplete without considering nuclear activation techniques which are still important analytical means in spite of the increasing use of optical, mass spectrometric and other methods that are independent of large facilities. In particular, thermal neutron activation analysis (NAA), preferably applied in the purely instrumental version (INAA), is indispensable as a reference method in view of the manifold opportunities and of the minor danger of losses and contamination.

For some matrices such as soil and sediments INAA is the method of first choice in spite of a rather long turnover time [7]. A distinct indication of this assessment is, for instance, provided by the IAEA/Soil-7 Intercomparison [42] in which 56 laboratories participitated with a variety of analytical methods. The first evidence is that the percentage of outliers was lowest for NAA in this intercomparison. Secondly, if 5 to 0 points are awarded to the participating laboratories in sequence of increasing deviation of the reported value from the overall mean for each element detected, the first eleven ranks of the resulting order of precedence are taken by NAA. This is shown in Fig. 5.

The development of high-intense neutron facilities such as KORONA [43—45] has opened new possibilities for 14-MeV NAA. A typical example is the mineralogical characterization of sediment samples [1]. The application in soil analysis is promising as well (cf. Fig. 5).

Numerous reviews on NAA may be found in the literature. The most recent ones with respect to environmental research were given by *Alian* and *Sansoni* [46] and, at the Mayaguez Conference [47], by *DeCorte* and *Hoste* [48], *Guinn* [49], and *Steinnes* [50]. Intercomparisons with ICP and TXRF were published by *Michaelis* et al. [6] and

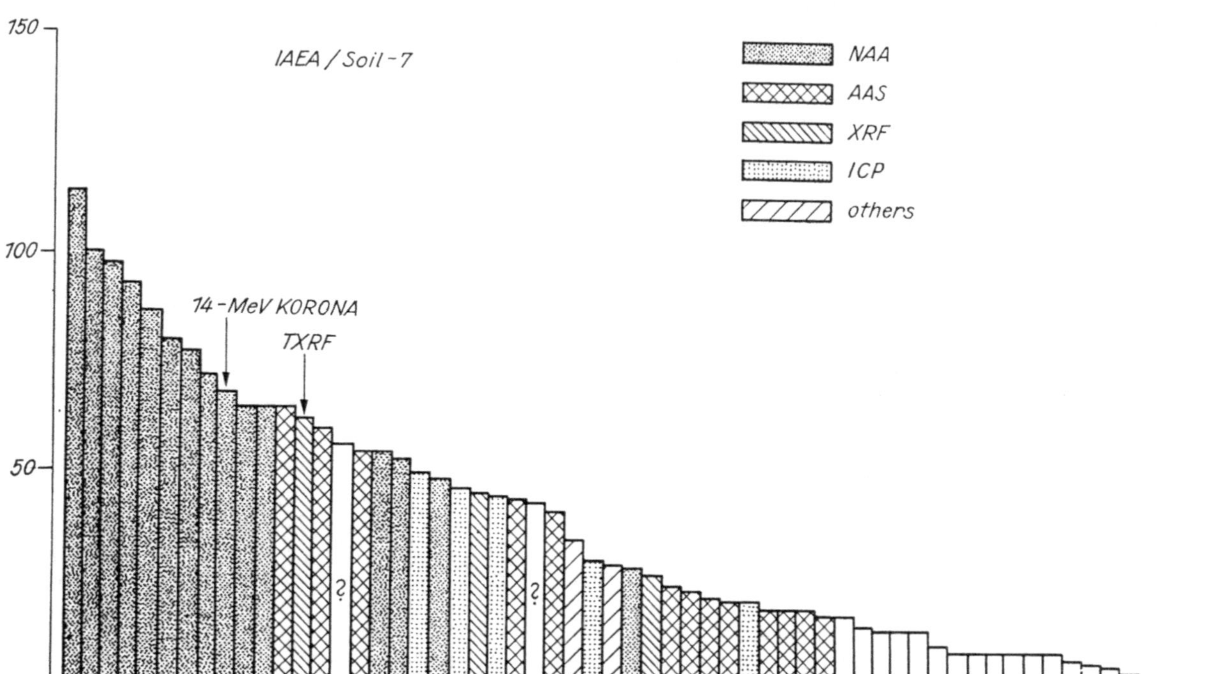

Fig. 5. Order of precedence of analytical methods in the IAEA/Soil-7 intercomparison. Evaluation based on data reported in [42]

Michaelis [7]. There seems to be an increasing trend to combine NAA with pre-irradiation chemical separation techniques, in particular in connection with speciation studies.

8. Trace gas analysis by ionization detectors

Ionization detectors take advantage of the dependence of the ionization current in an electrical field upon the composition of a gas mixture and the concentrations of the components, the gas volume being irradiated by an alpha or beta particle emitting radioactive source. Depending on the field strength, the detectors are operated in the recombination, saturation or multiplication mode. Correspondingly, they may be classified as electron capture and aerosol ionization detectors (ECD, AID), cross-section detectors (CSD), and noble-gas detectors (NGD), respectively. Suitable radiation sources are ^{3}H, ^{55}Fe, ^{63}Ni, ^{85}Kr, ^{90}Sr, and ^{241}Am. A detailed review on the techniques and possible applications has been given by *Leonhardt* [51].

ECDs are mainly used in gas chromatography to detect organic compounds (halocarbons, pesticides, herbicides, insecticides, metal organic compounds), but also inorganic gases such as SO_2 and N_2O. AIDs are able to work as gas analysers without any separation columns by converting the compound to be detected into aerosols using pyrolysis, gas-solid reactions, gas-gas reactions or chemical prereactions (*Leonhardt* [51], *Grosse* et al. [52]). The aerosols enhance the recombination rate which results in a change of the current proportional to the gas concentration. Various gases have been detected the most important being SO_2, NO_x, NH_3 and several halocarbons. The main advantage of AIDs lies in the possibility to design multipurpose devices which are adaptable to various measuring problems. The detection limits are between 1 and 40 ppb.

While CSDs have so far not reached great importance, NGDs have been used in gas chromatographic devices to detect a wide spectrum of compounds with detection limits in the ppb range.

9. Tracking of mud fields in tidal rivers

Mud fields above the consolidated sediment are supposed to contribute to a non-negligible extent to the pollutant transport in tidal rivers since they may reach thicknesses of several metres and since the pollutant contents of suspended particulate matter and sediments are quite high [53, 54]. The mud density usually varies between 1·0 and 1·4 g/cm³. This range is well accessible to radiometric density measurements if the energy of the gamma-rays is properly chosen. ^{137}Cs is the radionuclide which is commonly used for these purposes. Other techniques such as double-frequency echo sounding often fail owing to a variety of physical causes.

Fanger and coauthors [55, 56] have developed a radiometric density gauge which has proved to be a useful means for studying the three-dimensional distribution of mud fields and their temporal variabilities in tidal rivers and harbours. The set-up of the lowerable device is shown in Fig. 6. The system is operated from board a ship. It penetrates through the mud into the sediment by gravity. Besides the ^{137}Cs gamma-ray source and the ruggedized NaI(Tl) detector with a 25 cm transmission length, a piezoelectric pressure sensor and a biaxial inclinometer are incorporated in the device to measure the instantaneous diving depth and the inclination of the probe. Every tenth of a second the counting rate is converted into density values by means of a microprocessor. A typical density profile showing a gradual increase markedly above the "water" sediment interface is presented in Fig. 7. Also included is the recording of the inclinometer. As an example of application, the variation of three density horizons with time at one of several locations investigated in the Lower Weser River is plotted in Fig. 8.

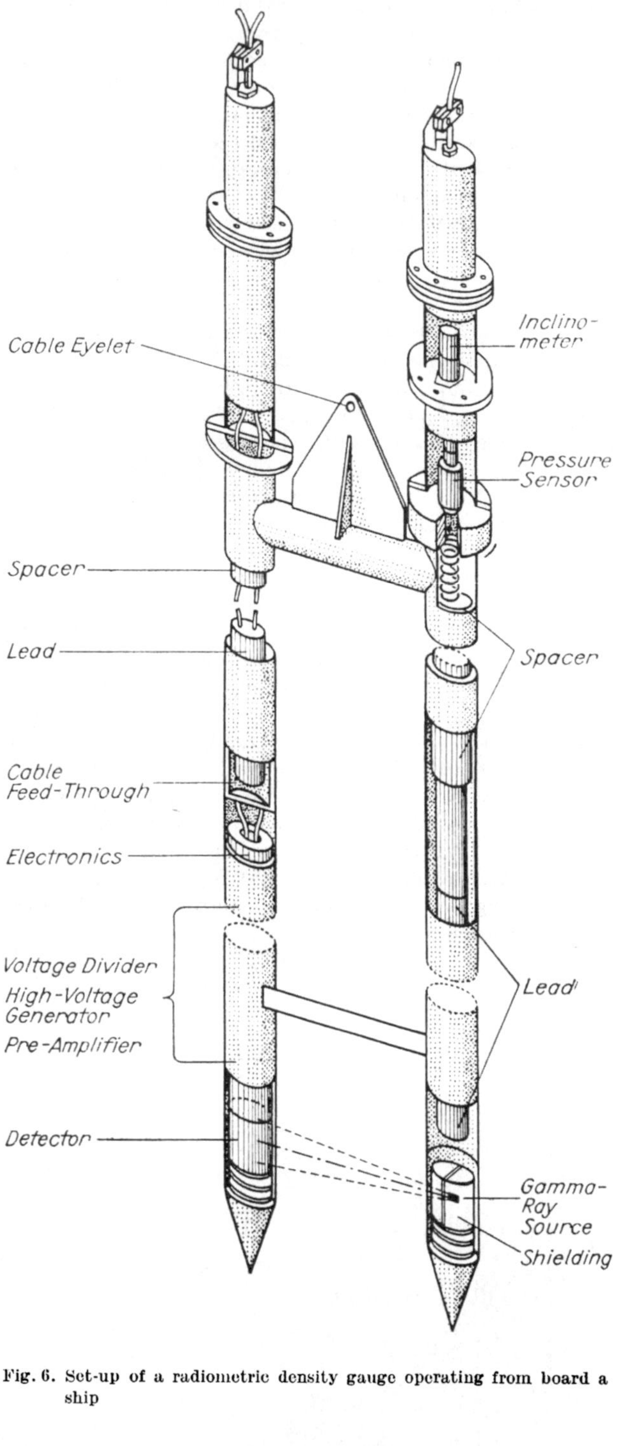

Fig. 6. Set-up of a radiometric density gauge operating from board a ship

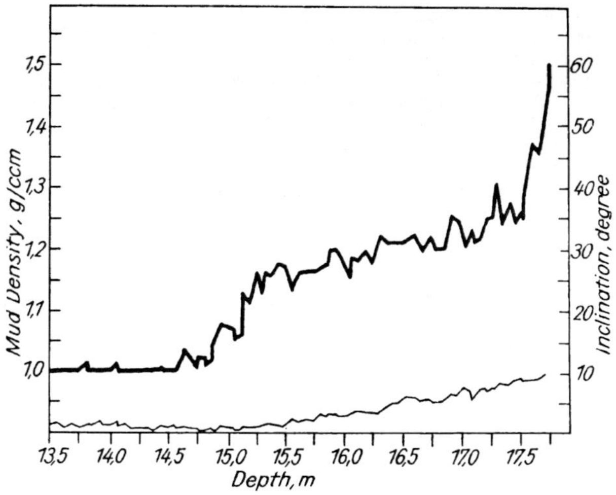

Fig. 7. Density profile of a mud field

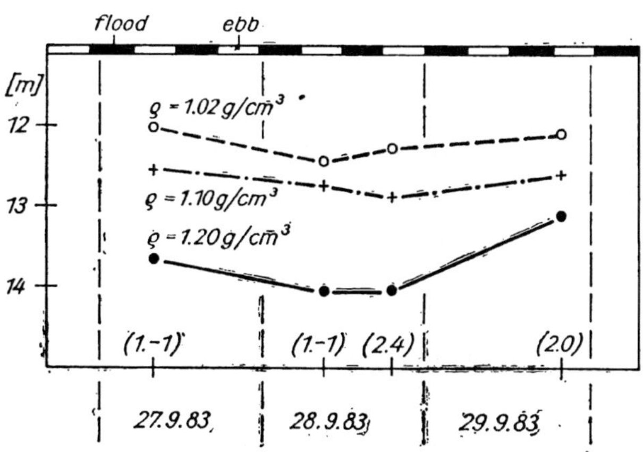

Fig. 8. Variation of three density horizons with time at one of the locations investigated in the Lower Weser River

10. Conclusions

The present paper exemplifies the use of radioisotopes in environmental research and protection by selecting a variety of fields of application. The purpose was to illustrate the considerable potential rather than to give a complete survey. The fields were selected according to the characteristic properties of which advantage is taken in radionuclide techniques:

(1) the occurrence of natural radioactivity,
(2) the possibility to label materials, compounds or elements,
(3) the high specificity of the emitted radiation,
(4) the high detection power for radioactivity,
(5) the occurrence of ionization effects, and
(6) the attenuation of nuclear radiation by materials.

Most of the problems described can be tackled with radioisotopes only, or radionuclide techniques are at least superior to non-nuclear methods.

Received September 27, 1985

References

[1] *B. Anders, W. Junge, J. Knoth, W. Michaelis, R. Pepelnik, H. Schwenke,* Proc. of the 5th Int. Conf. on Nuclear Methods in Environmental and Energy Research, April 2–6, 1984, Mayaguez, Puerto Rico; CONF-840408 (DE 84017348), 202–215

[2] *M. Spychala,* Thesis, University of Hamburg, 1986

[3] *M. Schönburg, W. Berneike, C. Weitkamp, W. Michaelis,* 3. Colloquium Atomspektrometrische Spurenanalytik, 18.–21. März 1985, Konstanz; Weinheim: VCH Verlagsgesellschaft (in press)

[4] *W. Michaelis, J. Knoth, A. Prange, H. Schwenke,* 33rd Annual Denver X-Ray Conference, July 30–August 3, Denver, CO, USA; Advances in X-Ray Analysis, Vol. 28, New York–London: Plenum Press 1985, 75–83

[5] *W. Michaelis, A. Prange, J. Knoth,* Symposium on Instrumental Multielement Analysis, April 2–5, 1984, Jülich, F.R.G.; in: *B. Sansoni* (ed.) Instrumentelle Multielement-Analyse, Weinheim: VCH Verlagsgesellschaft 1985, 269–290

[6] *W. Michaelis, H.-U. Fanger, R. Niedergesäß, H. Schwenke,* Symposium on Instrumental Multielement Analysis, April 2–5, 1984, Jülich, F.R.G.; in: *B. Sasoni* (ed.) Instrumentelle Multielement-Analyse, Weinheim: VCH Verlagsgesellschaft 1985, 693–710

[7] *W. Michaelis,* Colloquium Spectroscopicum Internationale XXIV, September 15–20, 1985, Garmisch-Partenkirchen, F.R.G.; Fresenius Z. Anal. Chem. (in press)

[8] *V. A. Fassel,* Symposium on Instrumental Multielement Analysis, April 2–5, 1984, Jülich, F.R.G.; in: *B. Sansoni* (ed.) Instrumentelle Multielementanalyse, Weinheim: VCH Verlagsgesellschaft 1985

[9] *Lal D. Krishnaswami, J. M. Martin, M. Meybeck,* Earth Planet. Sci. Lett. **11** (1971) 407–414

[10] *J. A. Robbins, D. N. Edgington,* in: *J. S. Marshall* (ed.) Proc. 2nd Federal Conf. Great Lakes (1976) 378–390

[11] *J. Dominik, U. Förstner, A. Mangini, H.-E. Reineck,* Senckenbergiana marit. **10** (4/6) (1978) 213–227

[12] *J. A. Robbins, D. N. Edgington,* Geochimica et Cosmochimica Acta **39** (1975) 285–304

[13] *I. L. Larsen, N. H. Cutshall, C. R. Olsen,* Proc. of the 5th Int. Conf. on Nuclear Methods in Environmental and Energy Research, April 2–6, 1984, Mayaguez, Puerto Rico; CONF-840408 (DE 84017348), 357–363

[14] *R.-P. Stößel, W. Michaelis, A. Prange,* Prod. Int. Conf. Heavy Metals in the Environment, September 10–13, 1985, Athens, Greece, Vol. 1, 203–205

[15] *R.-P. Stößel, A. Prange,* Anal. Chem. 57 (1985) 2880

[16] *R. Jonas, K. J. Vogt,* Jül-1780, April 1982

[17] *R. Jonas,* Jül-1949, September 1984

[18] *R. Jonas,* in: Luftreinhaltung – Luftverschmutzung, Tagung der Arbeitsgemeinschaft der Großforschungseinrichtungen (AGF), 3./4. November 1983, Bonn–Bad Godesberg, 24–26

[19] *K. Naumann,* Thesis, University of Hamburg, 1983

[20] *B. Georgi, W. Kühn,* in: Luftreinhaltung – Luftverschmutzung, Tagung der Arbeitsgemeinschaft der Großforschungseinrichtungen (AGF), 3./4. November 1983, Bonn – Bad Godesberg, 27–29

[21] *F. Führ,* Vorträge der 35. Hochschultagung, Landw. Fak. Univ. Bonn, 2./3. März 1982, Münster: Landwirtschaftsverlag Münster 1982, 115–122

[22] *F. Führ, W. Steffens,* in: Strahlung und Radionuklide in der Umwelt, Tagung der Arbeitsgemeinschaft der Großforschungseinrichtungen (AGF), 8./9. November 1984, Bonn–Bad Godesberg, 51–53

[23] *J. Wieneke, F. Führ,* in: *W. P. Duncan* and *A. B. Susan* (eds.) Synthesis and Applications of Isotopically Labelled Compounds, Proc. Int. Symp., Kansas City, MO, USA, June 6–11, 1982, Amsterdam: Elsevier Scientific Publishing Company 1983, 373 to 374

[24] *F. Führ, J. Wieneke,* Angew. Botanik **47** (1973) 97–106

[25] *P. N. Moza, I. Scheunert, W. Klein, F. Korte,* J. Agric. Food Chem. **27/5** (1979) 1120–1124

[26] *P. N. Moza, I. Weisgerber, W. Klein,* J. Agric. Food Chem. **24/4** (1976) 881–885

[27] *P. N. Moza, I. Scheunert, F. Korte,* Arch. Environm. Contam. Toxicol. 8 (1979) 183–189

[28] *R. Kloskowski, I. Scheunert, W. Klein, F. Korte,* Chemosphere **10/10** (1981) 1089–1100

[29] *I. Scheunert, J. Kohli, R. Kaul, W. Klein,* Ecotoxicology and Environmental Safety **1** (1977) 365–385

[30] *D. Freitag, I. Scheunert, W. Klein, F. Korte,* J. Agric. Food Chem. **32** (1984) 203–207

[31] *T.-S. Hsu, R. Bartha,* J. Agric. Food Chem. **24/1** (1976) 118–122

[32] *S. Jahnke, G. Stöcklin, J. Willenbrink,* Planta **153** (1981) 56–63

[33] *J. Wieneke, R. Fritz, G. Roeb,* in: Strahlung und Radionuklide in der Umwelt, Tagung der Arbeitsgemeinschaft der Großforschungseinrichtungen (AGF), 8./9. November 1984, Bonn–Bad Godesberg, 45–47

[34] *H. Moser, W. Rauert,* Isotopenmethoden in der Hydrologie, Stuttgart: Bornträger 1980

[35] Guidebook on Nuclear Techniques in Hydrology, 1983 edition, International Atomic Energy Agency, Vienna, Technical Reports Series No. 91

[36] *W. Drost, D. Klotz*, see [35], 223 – 256
[37] *V. I. Ferronsky, J. A. Czubek*, see [35], 257 – 271
[38] *C. Schnier, H.-U. Fanger*, GKSS 83/E/67
[39] *A. Prange*, Thesis, University of Hamburg, 1983
[40] *A. Prange, A. Knöchel, W. Michaelis*, Anal. Chim. Acta 172 (1985) 79
[41] *R.-P. Stößel, A. Prange*, Anal. Chem. 57 (1985) 2880
[42] *L. Pszonicki, A. N. Hanna, O. Suschny*, IAEA/RL/112 (1984)
[43] *H.-U. Fanger, R. Pepelnik, W. Michaelis*, Proc. of the 4th Int. Conf. on Nuclear Methods in Environmental and Energy Research, April 14 – 17, 1980, Columbia, MO, USA; CONF-800433, 195 – 204
[44] *H.-U. Fanger, R. Pepelnik, W. Michaelis*, J. Radioanal. Chem. 61 (1981) 147 – 163
[45] *R. Pepelnik, H.-U. Fanger, W. Michaelis, B. Anders*, J. Radioanal. Chem. 72 (1982) 393 – 403
[46] *A. Alian, B. Sansoni*, J. Radioanal. Nucl. Chem. 89 (1) (1985) 191 – 275
[47] Proc. of the 5th Int. Conf. on Nuclear Methods in Environmental and Energy Research, April 2 – 6, 1984, Mayaguez, Puerto Rico; CONF-840408 (DE 84017348)
[48] *F. DeCorte, J. Hoste*, see [47], 2 – 22
[49] *V. P. Guinn*, see [47], 23 – 42
[50] *E. Steinnes*, see [47], 135 – 140
[51] *J. W. Leonhardt*, in: Industrial Application of Radioisotopes and Radiation Technology, Conference Proceedings, Grenoble, 28 September – 2 October, 1981; IAEA, Vienna, 1982, 163 – 178
[52] *H. J. Grosse, J. W. Leonhardt, H.-R. Doering, F. Gleisberg*, in: Industrial Application of Radioisotopes and Radiation Technology, Conference Proceedings, Grenoble, 28 September – 2 October, 1981; IAEA, Vienna, 1982, 490 – 491
[53] *W. Michaelis*, International Conference on Heavy Metals in the Environment, September 6 – 9, 1983, Heidelberg, Vol. 2, 972 to 975; GKSS 83/E/43
[54] *W. Michaelis, H.-D. Knauth* (eds.): Das Bilanzierungsexperiment 1982 (BILEX '82) auf der Unterelbe, GKSS 85/E/3
[55] *H.-U. Fanger, W. Michaelis*, in: Strahlung und Radionuklide in der Umwelt, Tagung der Arbeitsgemeinschaft der Großforschungseinrichtungen (AGF), 8./9. November 1984, Bonn – Bad Godesberg: 35 – 38
[56] *H.-U. Fanger, E. Bössow, H. Kuhn*, Die Küste, No. 42 (1985) 189 – 207

Isotopenpraxis 22 (1986) 10, S. 344 bis 347

Principles of Constructing Set of Equipment and Realization of Software for Analysing of α-autoradiographic Images*)

I. A. Golenishchev, S. N. Grachev, I. G. Isaev, A. A. Kuz'min, V. M. Lyubakov, A. N. Mayorov

(State Committee for Uses for Atomic Energy, Moscow)[1])

The main peculiarities of constructing set of equipment for analysis of α-autoradiographic images are discussed. Hardware, software and metrological means are analysed. The structure of the equipment is presented providing for inspecting homogeneity of plutonium dioxide distribution by means of α-autoradiography under shop conditions. The software has been written on the basis of "RAFOS" (RT-II) operational system. The basic programming language is "FORTRAN IV". The necessary components and structure of application programs are discussed. As far as metrological means are concerned, built-in and application devices providing for control of the equipment operating characteristics stability are described.

Die wichtigsten Besonderheiten für den Aufbau eines α-autoradiographischen Bildnachweisgerätes werden diskutiert. Die Hardware, Software und die metrologischen Voraussetzungen werden analysiert. Die Versuchsanlage wird zum Nachweis der Homogenität einer PuO_2-Verteilung für die α-Autoradiographie unter Betriebsbedingungen vorgestellt. Die Software wurde auf der Basis von RAFOS (RT-II) geschrieben; das Grundprogramm ist in der FORTRAN-IV-Sprache. Notwendige Komponenten und die Struktur des Anwendungsprogramms werden diskutiert. Soweit es die metrologischen Teile betrifft, die eingebaut sind und zur Kontrolle der Apparatur dienen, so werden ihr Betriebsverhalten und ihre Stabilität beschrieben.

Keywords

alpha particles; autoradiography; fortran; images; plutonium dioxide; programming languages

The analysis of distribution homogeneity of nuclear fuel components as well as obtaining objective numerical magnitudes of this parameter is of primary importance for fuel element manufacturing. Usually the homogeneity of plutonium dioxide distribution is controlled by means of autoradiography. The procedure includes the comparison of enlarged α-autoradiogram images of pellets to the autoradiogram images of pellets certified in a conventional way [1]. The above-mentioned testing procedure provides for obtaining only qualitative and subjective data characterizing the plutonium dioxide distribution in the fuel.

The most promising means of obtaining quantitative data of homogeneity distribution is automated decoding of α-autoradiogram image by means of optico-electronics [2].

The introduction of automated decoding, providing for objective quantitative evaluation of distribution necessitates the quantitative evaluation of accuracy of reproduction of this characteristic on the autoradiogram. It also necessitates image resolution and contrast stabilization as well as defines the peculiarities of software and analyzing equipment construction [3].

Building block principle providing for the possibility of the structure rearrangement and program compatibility when microcomputer is applied is the basis of the analyzing

*) Presented at the XIVth International Symposium on "Autoradiography" November 19 – 23, 1984, Reinhardsbrunn, G.D.R.

[1]) Address: SU-109180 Moscow, Staromonetny per. 26

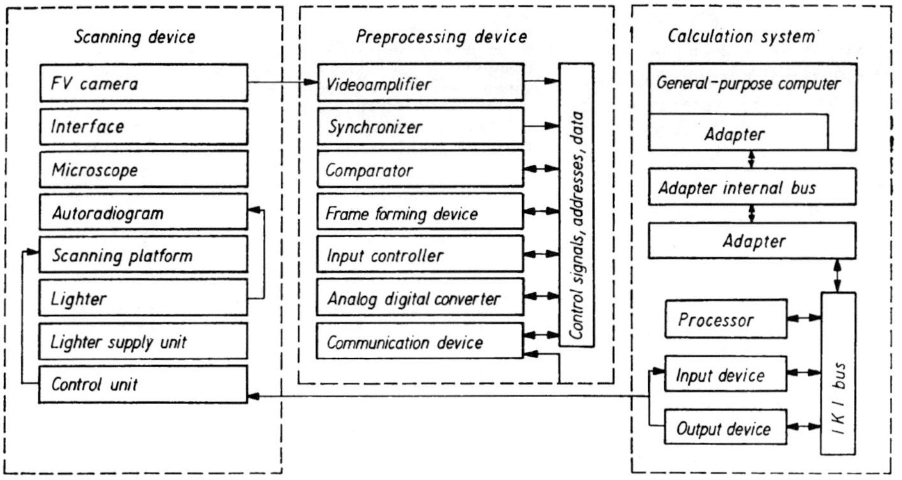

Fig. 1. General block diagram of a set of equipment for analysis of α-autoradiographic images

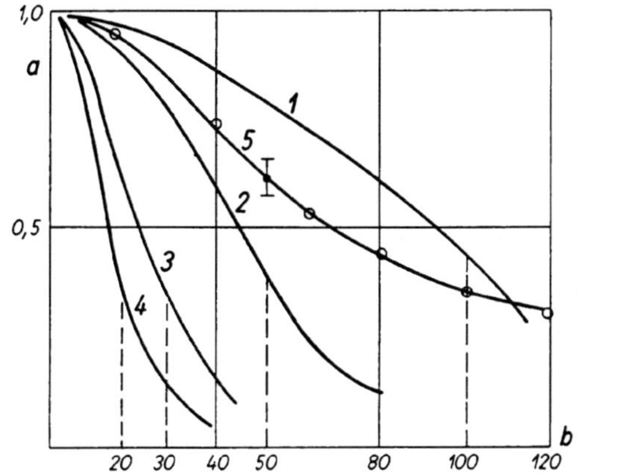

Fig. 2. The frequency-contrast characteristics of forming images and optico-electron channel.

a) frequency-contrast characteristics, M; b) spatial frequency, mm^{-1}

1 – for region of existence of the edge-spread function at 2 V = 10 μm;
2 – for region of existence of the edge-spread function at 2 V = 20 μm;
3 – for region of existence of the edge-spread function at 2 V = 40 μm;
4 – for region of existence of the edge-spread function at 2 V = 60 μm;
5 – the frequency-contrast characteristics of the optico-electron channel

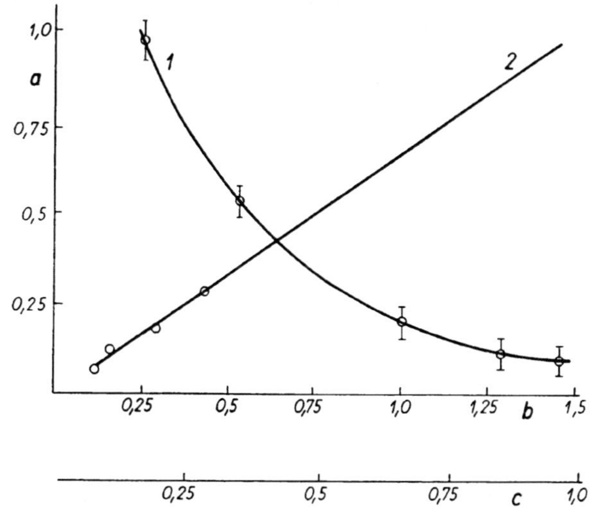

Fig. 3. The relationship of the videosignal voltage to the value of image optical density (1) and to the value of luminous flux (2);

a – videosignal voltage (V);
b – image optical density (D);
c – Φ/Φ_{max} is the normalized value of the luminous flux, where Φ_{max} is the luminous flux value corresponding to the minimum optical density of an autoradiographic image

equipment construction. The analyzing equipment consists of a scanning device; a data preprocessing device and a calculator (that is, calculation system) (Fig. 1).

The scanning device performs scaling and image-to-electric signal conversion. The motion from picture to picture is carried out according to a randomly choiced path by means of a scanning platform; the picture dimension is 1 mm^2 in the object plane. The correspondence of the scanning device parameters to the necessary characteristics of α-autoradiogram image transmission and conversion is established on the basis of analysis of edge-spread functions of α-autoradiogram imaging and frequency-contrast characteristic of the optico-electron channel.

Proceeding from the comparison of the frequency-contrast characteristic of α-autoradiogram imaging for various regions of the edge-spread function in Fig. 2 (1, 2, 3, 4) and the frequency-contrast characteristic of the optico-electron channel, it becomes evident that feeding of the signal along "object"–"electrical signal" through channel may be carried out practically without distortions for regions of the edge-spread function in the range of 2 V = 20 + 60 μm. Simultaneously, variation of image optical density (D) should be provided in the range of 0.3 + 1·2 that corresponds to the linear part of transfer characteristic of a TV camera (Fig. 3).

Thus, providing the above-mentioned characteristics, the quantitative interpretation of distribution of nuclear fuel components becomes possible in the range of concentrations from 0 to 100% for plutonium dioxide.

The determination of distribution value is provided with subsequent videosignal processing; its converting to digital code and making calculations according to the present algorithm [3].

The preprocessing device performs analog conversion of the signal for its subsequent amplitude normalization; destabilizing factor compensations and time conversion for subsequent analogue-to-digital conversion and quantization as well as for extraction of synchronizing signals. The digitizing of videosignal is provided along the lines with the frequency being equal to 10 mHz; and the quantization is provided using 256 levels for subsequent calculations and an arbitrary threshold for subsequent visual display.

The general-purpose mini-or-microcomputer-based calculator processes the converted videosignal in accordance with the present algorithm and operates the preprocessing device.

The preprocessing device (Fig. 1) is constructed on the basis of trunk-modularity that, in turn, is based on the use of functional building blocks and the trunk. The trunk line of the preprocessing device is a set of buses transmitting control signals, addresses and data. Such arrangement enables simultaneous operaion of separate blocks controlled by the calculator supplied with the communication device.

The calculator is based on the general-purpose computer and a set of facilities relating to local data-control systems (KTS LIUS-2). Three buses provide for easy assembling of a required calculator configuration and for high non-interaction degree of separate components of the system. On the basis of IK1 bus (interface channel 1 bus), i.e. KTS LIUS-2 in-crate trunk, the communication device is arranged. Input/output devices for digital signals; immediate-access memory; semipermanent memory; processor; energy-independent immediate-access memory and displays are applied as modules connected to IK1. By means of the special adapter the IK1 bus is connected to 1 + 4 adapter internal buses that will be considered IK1 extension. Up to 7IK1 buses and mini-computer SM4 bus or microcomputer "Elektronika 60" bus may be connected to the adapter internal bus by way of the adapters. Thus, the communication device and the general-purpose computer that are the parts of calculator become hardware-independent. The software is loaded into the semipermanent memory. Because of the restricted size of the immediate-access memory optical density data is processed in consecutive order, column after column, within the frames of the picture. The results of column and picture processing are stored in the immediate-access memory of the computer or provided it is necessary, the results are stored in the energy-independent immediate-access memory of KTS LIUS-2. The programs are functionally complete. The software is written in FORTRAN IV; the drivers are in assembler language in RAFOS (RT-11) operational system.

Lift of programs and their basic functions are represented in Tab. 1.

Tab. 1.

Program name	Short description of the program functions
1. POTL	The program determines the relative portion of area and the mean dimension of vertical chord in the picture regions having optical density within the predetermined ranges.
2. DIAGON	The program determines for black-and-white images: 1) distribution of blackening regions according to dimensions and diameters of equivalent circles; 2) relative area of blackening regions; 3) relative area of blackening regions with the equivalent circle diameter exceeding the predetermined value.
3. DIAGN	The program operation is similar to DIAGON program, but the initial image is considered a "gray" one quantified by the number of levels not more than 5.
4. FON	The program eliminates the background arising in the initial image as a result of instability of the videosignal amplitude.
5. GRAN	The program determines the coordinates of secants of blackening regions for each line and level.
6. SUM	The program defines the relation of different secants to the same region within the whole picture and for each digitization level.
7. MGU 3	The program determines the area, the maximum dimension and the diameter of an equivalent circle provided the coordinates of its secants are known.
8. GIST	The program provides for printing the histograms of the region distribution according to the maximum dimensions and diameters of equivalent circles.
9. DRVD	The program provides for loading the data characterizing the optical density of the picture line into the immediate-access memory of the computer.

Tab. 2

Names of images	Short descriptions of characteristics and features	Purpose
1. OMP	Objective micrometer of transmitted light	Determination of dimensions and gain factor
2. MOV1-15HU42	Eyepiece micrometer	Determination of dimensions and gain factor
3. Set of test objects	5 + 200 lines/mm	Determination of frequency-contrast characteristic
4. Optical wedge	Range of measuring optical density: from 0·2 to 1·5	Determination of transfer characteristic
5. Test-object 1	A slide containing the image that shows points and background having different optical densities in the range of 0·2 to 1·3 units	Test of contrast sensitivity
6. Test-object 2	A slide containing the image of geometrical figures with the predetermined dimensions	Determination of measurement errors provided linear dimensions are measured
7. Test-object 3	An α-autoradiogram with the certified dark-to-bright field ratio on different levels of optical density	Determination of measurement errors provided a relative portion of area is measured

POTL, DIAGON and DIAG are master programs independent of each other; the rest ones are subroutines that are used in the master programs.

The metrological means are essential for providing control of status and stability of the analyzing equipment characteristics. The metrological means of the set equipment are based on applied test-objects and simulators of test-objects produced by the equipment. List of basic test-objects is given in Tab. 2.

The hardware-controlled simulators of test-images generate a videosignal at the input of the system bearing information of the image with the definite parameters. The formed image presents a regular structure (square field, vertical or horizontal strips) with the present parameters [6]. In the equipment under consideration a vertical strip field is applied.

Basic specifications of the equipment

1) Scanning (interlaced scanning)
   - number of lines — 625
   - number of pictures — 25
   - screen size — 4:3
   - camera tube — vidicon LI 451
2) Maximum area of the image portion analyzed — 1 mm × 1 mm
3) Range of variation of image optical density — $0{\cdot}2 + 1{\cdot}3D$
4) Digitization frequency — 10 mHz
5) Number of quantization levels — 256
6) Minimum frame size — 1 × 1
   Maximum frame size — 512 × 512
7) In automatic mode the equipment provides:
   - to plot the histogram of distribution of particle images in accordance with the dimensions of maximum chords and diameters of equivalent circles;
   - to determine the relative portion of area occupied with the image of the particles;

– to determine the mean value of the relative portion of area if several pictures are taken into account.

8) Voltage supply
220 V single-phase alternating current; 50 Hz frequency.

Received October 19, 1985

References

[1] *E. A. Aitken, S. K. Evant*, Transactions American Nuclear Society 13 (1970) 602

[2] *V. V. Yampol'sky, A. N. Mayorov, L. S. Nosukhin* et al., Rising contrast of autoradiogram half-tone images, Proceedings of the 13th International Symposium on Autoradiography in Tábor, Czechoslovakia (May 3 – 5, 1983), 70 (in Russian)

[3] *I. A. Golenishchev, A. V. Grachev, I. G. Isaev* et al., Autoradiographic Investigations of Distribution of Nuclear Fuel Components, Proceedings on the 13th International Symposium on Autoradiography in Tábor, Czechoslovakia (May 3 – 5, 1983), 96 (inRussian)

[4] *N. N. Krasil'nikov*, Statistical Theory of Image Transmission, "Svyaz", Moscow, 1976 (in Russian)

[5] *J. van Dsele*, Television Scanner. Automation and Inspection Applications of Image Processing Techniques, London, Sept. 12 – 13, 1977 Proc. Soc. Photo Opt. Instrum., Eng. 1977

[6] *I. N. Gulin*, TV Image Electron Synthesis, Moscow: Sovetskoye radio, 1979 (in Russian)

Isotopenpraxis 22 (1986) 10, S. 347 bis 350

Formalization of Autoradiogram Image Forming Process on Dielectric Track Registration Detectors*)

A. A. Akulinin, A. V. Grachev, A. N. Mayorov, V. V. Yampol'sky

(All-Union Scientific Research Institute of Radiation Techniques)¹)

Some aspects of forming of α-autoradiogram images by dielectric track detectors made from cellulose nitrate are considered The nature of interaction of α-radiation and irradiation medium material and track registration detector material, as well as linear dependence between image optical density and the number of α-particles registered by the detector allow to understand the process of forming images to linear images and characterize it as a modulation transfer function. This mathematical formalism is applied to some examples which are considered in detail.

Es werden einige Gesichtspunkte bei der Entstehung von α-Autoradiogrammbildern betrachtet, wie sie bei Verwendung von Zellulosenitrat als einem dielektrischen Spurdetektor gebildet werden. Die Natur der Wechselwirkung von α-Strahlung und bestrahltem Material und dem Spurdetektor, wie auch die lineare Abhängigkeit zwischen der optischen Dichte des Bildes und der Zahl der registrierten α-Teilchen durch den Detektor kann als Vorgang der Bildentstehung so verstanden werden, als ob eine Modulation auf eine Übertragungsfunktion stattfinden würde. Dieser mathematische Formalismus wird an einigen Beispielen näher beschrieben.

Keywords

alpha particles; autoradiography; cellulose esters; dielectric track detector; fluctuations; plutonium dioxide; transfer functions

Wide use of dielectric track registration detectors for visualization of spatial distribution of radioactive elements emitting heavy charged particles [1] makes it necessary to develop a model which would provide for characterization of image reproduction quality on the autoradiogram. The detectors are also used for half-tone imaging which represents the full topology of activity distribution [2, 3]. Separate influence of imaging process parameters on spatial distribution and detectability of details of the image are investigated in a numer of papers [4, 5].

Mutual influence of blurring and optical density fluctuations caused by discrete nature of imaging on the image reproduction quality on the autoradiogram in the process of evaluation of mixed oxide fuel homogeneity is discussed.

Mutual influence of both factors, photographic and autoradiographic one as well, when analyzing static image forming systems may be characterized by information capacity [6].

The following representation of information capacity derived from *Shannon's* formula [7] for one-direction transmission in case of an isotope autoradiogram imaging system will be used:

$$I = \pi \int_0^{W_{\text{boundary}}} \log_2 \left[\frac{R(\nu) |M(\nu)|^2 K^2 + n(\nu)}{n(\nu)} \right] \nu \, d\nu \qquad (1)$$

where: $R(\nu)$, $n(\nu)$ – is the signal and noise power spectra respectively;
is normalized-per-unit frequency-contrast characteristic (FCC) of an autoradiogram imaging system;
W_{boundary} – is boundary frequency transmitted by an imaging system;
K – is coefficient characterizing the specific contrast.

Hence, provided values constituting expression (1) are determined, quantitative criteria may be obtained characterizing the reproduction of plutonium dioxide concentration distribution in a pellet on the autoradiogram.

The nature of interaction of α-radiation with the material of irradiating environment and the material of track registration detector as well as linear relation of image optical

*) Presented at the XIVth International Symposium on "Autoradiography", November, 19 – 23, 1984, Reinhardsbrunn, G.D.R.

¹) Address: SU-115230 Moscow, Warshavskoye Shosse 46

density to the number of tracks of α-particles registered by the detector provide for attributing the process of autoradiogram imaging on a dielectric track registration detector to linear mapping. The process may be investigated using a body of mathematics of *Fourier* transform. Previously, in paper [5] it was reported that the basic components of autoradiogram imaging process, namely, the formation of α-radiation field on the surface of an object under estimation; the formation of the latent image of the autoradiogram and the formation of the visible image of the autoradiogram have the triangular function of scattering. Consequently, the relation of the image detail contrast to their mutual disposition may be characterized by the following frequency-contrast characteristic:

$$M(\nu) = \frac{2}{(\pi\nu)^2\ u^2}(1 - \cos 2\pi\nu u) \qquad (2)$$

where: ν is the spatial frequency (mm^{-1});
u is half of the region of existence of edge-spread function.

The frequency-contrast characteristics of autoradiogram imaging on a dielectric track registration detector for various regions of existence of the edge-spread function characterized by the thickness of a compensating filter are shown in Fig. 1.

The analytical expression of frequency-contrast characteristics for autoradiogram imaging process makes it possible to determine the boundary frequency which may be transmitted in the process of imaging. It may be represented as

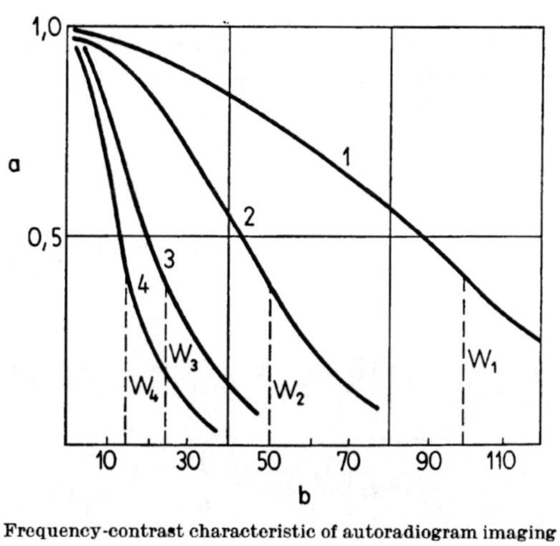

Fig. 1. Frequency-contrast characteristic of autoradiogram imaging process for various regions of edge-spread function
1 – $2U$ = 10 µm; 2 – $2U$ = 20 µm; 3 – $2U$ = 40 µm; 4 – $2U$ = 60 µm
a) frequency-contrast characteristic, $M(\nu)$; b) ν, frequency, mm^{-1}

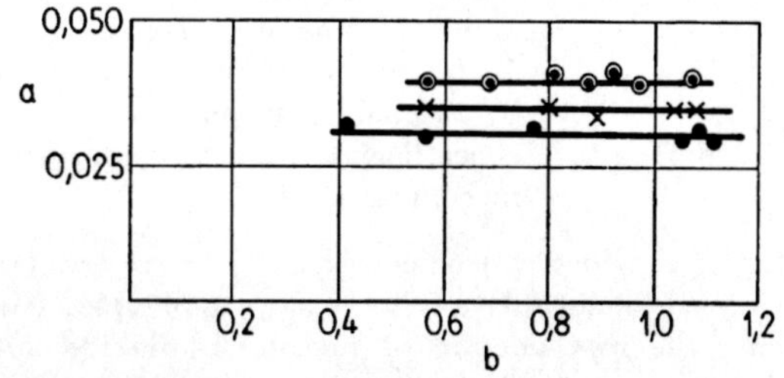

Fig. 2. The relationship of mean-square granularity to optical density for different diameters of tracks forming the image
1 – d = 1·5 µm; 2 – d = 2.0 µm; 3 – d = 2 µm
a) mean-square granularity, σ_D; b) optical density, D

follows:

$$W = \int_0^\infty M(\nu)\,d\nu = \frac{1}{2u} \qquad (3)$$

Provided the frequency characteristics of reproduction of plutonium dioxide concentration distribution on the autoradiogram are known, spectra of power of local changes in concentration distribution and fluctuations of optical density of the image should be studied. As a result of visualization of latent images, noise conditioned by the static nature of α-radiation which may be regarded as "white" noise [8] is added to noise caused by fluctuations of scattering capacity of separate etched tracks of α-particles.

In the process of investigation of spectra of fluctuation power of image optical density on track registration detectors made of cellulose nitrate it was concluded that in the frequency band which may be transmitted in the process of autoradiogram image formation power spectrum may be represented as "white" noise. It may be characterized by dispersion or by mean-square granularity. Plots in Fig. 2 show that for track registration detectors of cellulose nitrate mean-square granularity doesn't depend on optical density in optical density range corresponding to the linear part of the densitometric characteristic of the detectors; the density of the registered tracks ranges from 5×10^6 to 5×10^7 I/cm^2. This provides for increase of signal-to-noise ratio for autoradiogram image due to increase of optical density in the range corresponding to the linear part of the densitometric characteristic. For silver halide photographic materials signal-to-noise ratio peak is obtained if optical density corresponds to the beginning of the linear part of the characteristic curve [6].

Proceeding from plots similar to those shown in Fig. 2, and important conclusion can be drawn about the linear dependence of the mean-square granularity on the mean diameter of the etched tracks in the diameter range from 1 to 4 µm for α-particle track densities where overlap effect may be ignored. That provides to obtain the signal-to-noise ratios necessary for subsequent image interpretation by varying etching duration and exposure time. An autoradiogram image information capacity substantially depends on these ratios.

Before determining the spectrum of power of local inhomogeneities in plutonium dioxide concentration distribution (this spectrum simultaneously will be the input signal spectrum) the frequency-power relation should be determined for different, frequently occurring inhomogeneities. Proceeding from the kinetics of the procedure of mixed fuel pellet sintering and on the basis of experimental data, functional relations repesented in Tab. 1 (1st column) were chosen for approximation of local inhomogeneities of plutonium dioxide concentration distribution in case of different sintering modes. Frequency and spectral characteristics of approximation functions are given in columns II and III of the table, respectively. As the investigation showed, spectral characteristics of all functions drop steeply while the frequency rises, and up to 90% of power is concentrated in a narrow frequency range from 0 to 40 mm^{-1}. That provides for characterizing the spectrum of power of local changes in the concentration distribution of plutonium dioxide by the dispersion interpreted as a mean signal power with limited spectrum, according to the conclusions given in paper [9].

Consequently, taking into account all the above-mentioned data, formula (1) determining only the boundary of

Tab. 1. Local inhomogeneities in plutonium dioxide concentration distribution

Spatial presentation of local inhomogeneities $S(x)$	Frequency presentation of local inhomogeneities $S(\nu)$	Spectrum of power of local inhomogeneities $\|S(\nu)\|^2$	Dispersion $D_x = \delta_x^2 = \int_0^\infty \|S(\nu)\|^2 \, d\nu$
$S(x) = \begin{cases} 0 & \|x\| > \tau \\ A_0 & \|x\| < \tau \end{cases}$	$2A_0\tau \sin \frac{(2\pi\nu\tau)}{2\pi\nu\tau}$	$4A_0^2\tau^2 \frac{\sin^2 (2\pi\nu\tau)}{4\pi^2\tau^2\nu^2}$	$A_0^2\tau$
$S(x) = A_0 \exp(-x^2/\tau)$	$4A_0\sqrt{\pi\tau} \exp\left[-\frac{(2\pi\nu\tau)^2}{4}\right]$	$A_0^2\pi\tau^2 \exp\left[-\frac{(2\pi\nu\tau)^2}{2}\right]$	$7{\cdot}75A_0^2\tau$
$S(x) = \begin{cases} 0 & \|x\| > \tau \\ A/2\left[\cos\left(\frac{\pi x}{\tau}\right) + 1\right] & \|x\| \leqq \tau \end{cases}$	$\frac{A_0}{2\pi}\left[\frac{\sin 2\pi\tau\nu}{(1 - 4\tau^2\nu^2)\nu}\right]$	$\frac{A_0^2 \sin^2 (2\pi\tau\nu)}{4\pi^2\nu^2(1 - 4\tau^2\nu^2)^2}$	$3{\cdot}8A_0^2\tau$
$S(x) = \frac{2A_0}{e^{x/\tau} + e^{-x/\tau}}$	$A_0\pi\tau \sec h\,(\pi^2\tau\nu)$	$A^2\pi^2\tau^2 \sec h^2\,(\pi^2\tau\nu)$	$A_0^2\tau$
$S(x) = A_0 \exp\left(-\frac{\|x\|}{\tau}\right)$	$\frac{2A_0\tau}{(1 + 4\pi^2\nu^2\tau^2)}$	$\frac{4A_0^2\tau^2}{(1 + 4\pi^2\nu^2\tau^2)^2}$	$\frac{A_0^2\tau}{2}$

the information capacity for unit of autoradiogram area will look as follows:

$$J = \pi \int_0^{W_{\text{boundary}}} \log_2 \left[\frac{D_x[M(\nu)]^2 K^2}{\sigma^2} + 1\right] \nu \, d\nu \tag{4}$$

where: D_x — is the dispersion of local inhomogeneities of dioxide plutonium concentration distributions;

σ — is the mean-square granularity characterizing fluctuations of optical density in an autoradiogram image;

K — is the factor characterizing image specific contrast approximately equal to $0{\cdot}01 D/\%$ of plutonium dioxide concentration, the autoradiogram may be obtained with the mean spectrum of inhomogeneities in concentration distribution (that is, input signal) [6].

However, when calculating the information capacity of the autoradiogram image using formula (4) some peculiarities of image formation should be taken into account. These

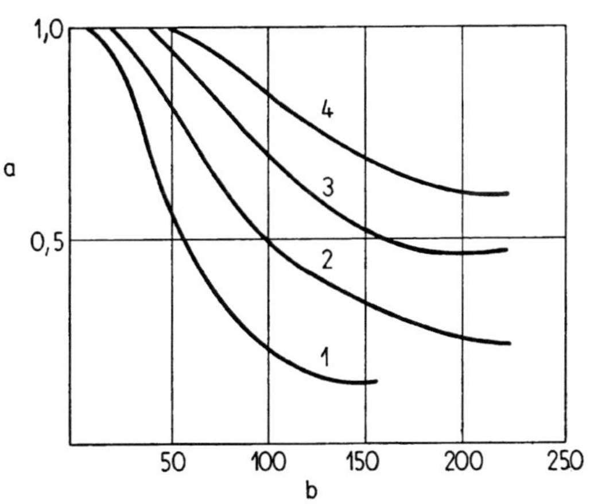

Fig. 3. The relation of autoradiogram image optical density (in relative units) to boundary frequency for collimation filter materials in which plutonium α-radiation path length is equal to

1 — 45 μm; 2 — 30 μm; 3 — 20 μm; 4 — 15 μm

a) optical density; b) boundary frequency, mm^{-1}

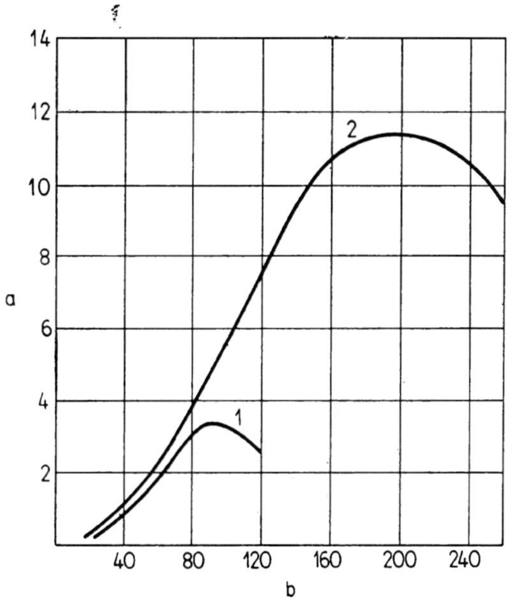

Fig. 4. The relation of the information capacity of the unit of the image area to the boundary frequency transmitted in the process of image formation, if materials with plutonium-emitted α-radiation path equal to

1 — 45 μm; 2 — 30 μm are used as collimating filters

a) information capacity, J · 10^{-4}, bit · mm^2; b) boundary frequency, mm^{-1}

peculiarities concern the relation of contribution of a single track in the resulting optical density to the residual energy of α-particles; the residual energy decreases as the collimation filter thickness is growing. The following factors were applied for obtaining plots represented in Fig. 3 the experimental data [10] showing simultaneous decrease of track scattering capacity and α-particle energy; the results of calculation of α-particle residual energy after transmission through the collimating filter providing for the definite region of existence of the edge-spread function for autoradiogram imaging as well as the transition from half-region of the edge-spread function existence to boundary frequency that may be transmitted in the process of image forming according to formula (3). The plots characterized the relation of optical density variation of the image to the boundary frequency growth caused by the growth of thickness of the collimating filter. These plots are represented in Fig. 3.

Taking into account the plots in Fig. 3 and using formula (4) the relations of the information capacity of autoradiogram image to the boundary frequency transmitted in the process of imaging were calculated. As an example, in Fig. 4 relationships for using various materials with different stopping power as collimation filters are given. Proceeding from analysis of these relationships it should be concluded that use of a material with essential stopping power as a collimating filter, provided the value of a ratio of the filter thickness to the α-radiation path length is the same, not only decreases the exposure time but also improves the image information capacity. The extreme on the plots provides for determination of the collimation filter optimum thickness that minimizes losses of information in the process of the autoradiogram imaging.

However, as the theoretical studies and experiments showed, increase of the collimating filter thickness up to the thickness value providing for the transmission of the boundary frequency, almost equal to the optimum one, in the process of imaging accounts for strict requirements to the reproducibility of the filter thickness and to the thickness range of the filters used. To provide for substantial reproducibility of the optical density the thickness shouldn't fluctuate by more than $\pm 0{\cdot}1$ μm. It is accounted for by the above-mentioned dependence of the contribution of a single track to the optical density of the autoradiogram image on the α-particle residual energy in the energy range less than 1 MeV.

It is not practically feasible to meet this requirement while autoradiographically inspecting radioactive components distribution under shop conditions. Therefore, for practical purposes, as far as obtaining reproducible results of analysis is concerned, especially in case of automatic quantitative data interpretation it would be senseless to increase the thickness of the collimation filter up to more than 0·9 α-radiation path in the filter material. And to provide for the transmission of the necessary boundary frequency in the course of the autoradiogram imaging the choice of the filter material should be made taking into account the relationships presented in Fig. 5 as well as the technological properties of the material that provide the formation of low-thickness homogeneous foils or films

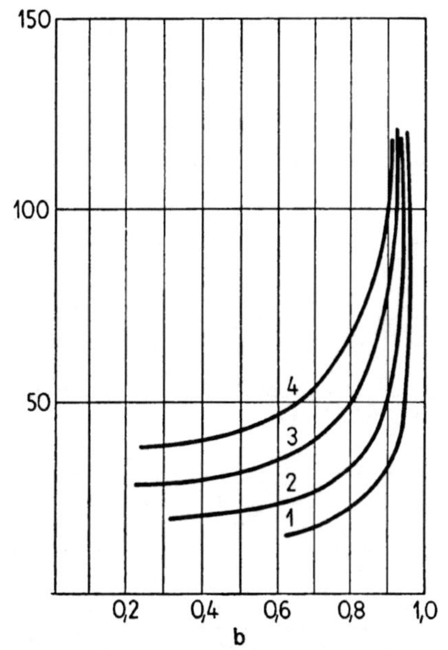

Fig. 5. The dependence of boundary frequency transmitted in the process of autoradiogram imaging on the ratio of the collimating filter thickness to α-particle path in the filter material. This dependence is plotted for materials with plutonium-emitted α-radiation path equal to

1 – 45 μm; 2 – 30 μm; 3 – 20 μm; 4 – 15 μm

a) boundary frequency, mm^{-1}; b) filter thickness/α-particle path length, *h/R*

Received October 19, 1985

References

[1] *R. L. Fleischner, P. B. Price, R. M. Walker*, Nuclear Tracks in Solids, Berkeley – Los Angeles – London: University of California Press 1975

[2] *L. Medveczky*, Isotopenpraxis **14** (1978) 185

[3] *W. I. Gruber*, Autoradiography of Irradiated Nuclear Ceramic Fuels, Proceedings of 17th conference on remote systems technology, 1969

[4] *R. Ilic, M. Najzer, A. Podgornik*, The Resolving Power of Autoradiography with SSNTD, Proceedings of XIth International Conference on Nuclear Photography and SSTD, 1978, 655

[5] *I. A. Golenishchev* et al., Autoradiographic Investigations of Distribution of Nuclear Fuel Components, Proceedings of 13th ISA, Tâbor, Czechoslovakia, May 1983, 96

[6] *H. Frizer*, Photographic Registration of Data, Moscow: Mir 1978

[7] *K. Shannon*, Mathematical Theory of Communication, M., IL, 1969

[8] *F. N. Novitsky, A. M. Yakobson*, Reduction of Random Noise with Slow Response Element of an I, Defectoscopy **4** (1970) 122 to 125

[9] *H. Kanamori*, Japanese Journal of Applied Physics **7** (1968) 414

[10] *J. Charalambous, M. Zamani*, Nuclear Instruments and Methods **138** (1976) 311

Isotopenpraxis 22 (1986) 10, S. 351 bis 353

Quantitative Interpretation of Autoradiograms with the Help of Determining Silver in them*)

A. G. Mironov, G. N. Zaguzin

(Geological Institute of the Bury'atian Branch of the Sibirian Department, Academy of Sciences of the USSR)[1])

The quantity of silver in districts of blackening of the autoradiograms is determined by means of the X-ray spectral microprobe analyzer MS-46 "Cameca". The experiments were made by applying ^{63}Ni. A calibration chart allows to determine the distribution coefficients of Ni in different phases and some examples show the usefulness of this suggested technique.

Die Menge an Silber in Schwärzungsbereichen von Autoradiogrammen wurde mit Hilfe einer Röntgenspektral-Mikrosonde MS-46 "Cameca" bestimmt. Die Versuche wurden mit dem Radionuklid ^{63}Ni ausgeführt. Ein Eichdiagramm erlaubt die Bestimmung von Verteilungskoeffizienten des Ni in unterschiedlichen Phasen, was an einigen Beispielen gezeigt wird, um die Brauchbarkeit der vorgeschlagenen Technik zu demonstrieren.

Keywords

autoradiography; measuring instruments; nickel 63; nondestructive analysis; phase studies; probes; X-ray emission analysis

Photometry methods based on measuring the degree of photo-emulsion blackening under the influence of radiation are traditionally used for the quantitative estimation of the non-uniformities of distribution of the elements with the help of autoradiography. In this case correlation, for example, of two phases is made by way of comparing the intensity of the light stream passing through the exposed districts. It is established that for the majority of radioisotopes in the interval of a blackening density from 0 up to 2 concentrations of the elements are proportional to the blackening density S,

$$C = kS\,,$$

where k is the coefficient of proportionality depending on the type and energy of radiation as well as the nature of the determined samples.

The method generally accepted and used in many fields of science and technology. However there are essential shortages, reducing the possibilities of microphotometry applications for the interpretation of autoradiograms. Namely, the dimension of blackening districts under investigation is limited; in most cases multiple, sometimes long exposures of the samples are necessary. Finally, during the study of the phases with widely different contents of the given element (the partition coefficient exceeding 100) the microphotometry is not effective at all due to the quick saturation of photoemulsion blackening above one of the phases.

There are other optical methods of quantitative estimation of the influence of the radioactivity on photoemulsion. They are based on calculating the quantity of α-tracks or silver grains manifested in a photoplate area unit. In this case the sensitivity increases. However optical calculation is limited by the resolving power of microscope and the necessity to use special very thin-layer (1 – 2 molecular layers) plates for the nuclear studies, missing in serial production.

Recently with the use of the electron microprobe analysis there appeared a possibility to determined silver quantity in the exposed districts of the photoemulsion rather than blackening density. We have improved the methods of quantitative interpretation of autoradiograms on the basis of the silver determination in the emulsion of the plates for the nuclear research in the electron microprobe analyzer MS-46 "Cameca". Serial plates of MA and MR with the thickness of the emulsion of 10 μm and the quantity of silver up to 82 – 85 wt.% in them have been used.

The experiments were conducted with the radioisotope ^{63}Ni (β-radiator with the average of β-particles equal to 65·9 keV). In the first place the investigation were aimed at revealing the possibility of the using home-made photoplates for measuring the silver content in them and building the calibration chart. Unlike the American investigators of the Carnegy Institute [1], who used one standard with different exposures, we exposed standard solutions with different contents of radioisotope ^{63}Ni. The standard solution were prepared on the basis of the original solution of nickel chloride by the activity of 5 mCi by way of its dilution in distilled water. The concentrations of ^{63}Ni constituted from 0·0004 up to 1·5 ppm. Then the drops of the standard solutions with a volume of 0·005 ml with the help of a microdose were directly deposited on the plates for the nuclear investigation (MA and MR) and were exposed during 10 days.

It should be pointed out that during drying up of the drop the matter contained in it can be distributed irregularly. This is especially vivid during the autoradiography, for example, of uranium in a drop of water according to the data of radiography. However, ad hoc the use of higher concentrations of the radioisotope and a gelatined layer of the photoplate allows to receive, during their developing, sufficiently similar spots of blackening with

*) Presented at the XIVth International Symposium on "Autoradiography" November, 19 – 23, 1984, Reinhardsbrunn, G.D.R.

[1]) Address: SU-670046 Ulan Ude, Pavlova 2

the variations in blackening density not more than 10 to 15%. The 13 pairs of drops with the contents of ^{63}Ni from 0·0004 up to 1·5 mg/l were deposited on a plate of 9 cm × 12 cm. After the plate was developed it was established that only a third for the drops can be quantitatively photometered; the other drops forming the spots with the blackening density over 2.

The measurement of the silver quantity in the exposed plates were carried out in the microanalyzer MS-46 "Cameca". The survey conditions: 30 kV accelerating potential, 20 nA regulated beam current, 10 μm × 50 μm scanned area, 20 s counting time. 20—40 measurements were made from each spot, a mean value and a standard deviation were counted. According to the measurements a calibration chart was drawn in which the contents of ^{63}Ni were in X-axis in a logarithmic scale and the contents of silver in Y-axis (Fig. 1).

The chart has a form of a typical curve with the districts of linear and non-linear relations in the field of low and high concentrations of ^{63}Ni. The chart shows distinctly that the field of linear relation is wide enough and covers the ^{63}Ni content range from 1 to 1,000. Supplementary detailing of the districts in the field of low and high concentrations of ^{63}Ni can be conducted though there is no practical necessity for this.

Backgrounds were measured on the unexposed areas of the emulsions. Metallic silver was used as a standard. Judging from the total concentration of silver in the photoemulsion, substantiated also by the measurement of its quantity in the completely spoilt part of the plate 24 impulses (24 imp/wt.%) correspond to one per cent of silver. In this manner a possibility of the use of silver concentrations in the emulsion of the plates for the nuclear investigations of the type MR and MA with the aim of quantitative interpretation of the autoradiograms is established quite definitely. Considerable advantages in comparison with the existing methods are established in this case.

1. The methods features higher resolution in comparison with the microphotometry as the probe diameter allows to analyse the blackening districts measuring 2—5 μm.
2. The range of the concentrations which can be determined with the help of autoradiography increased more than by two orders. This can be accounted by the facts that the saturation of emulsions in silver content is obtained much later than in their blackening density.

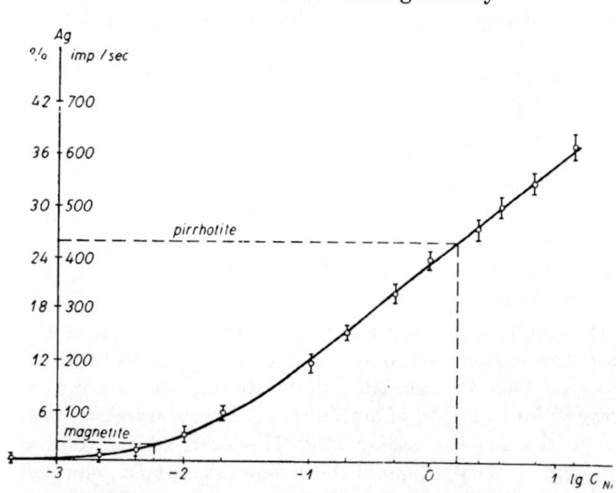

Fig. 1. The calibration chart of the independence of silver content in emulsion from the contents of ^{63}Ni in solution

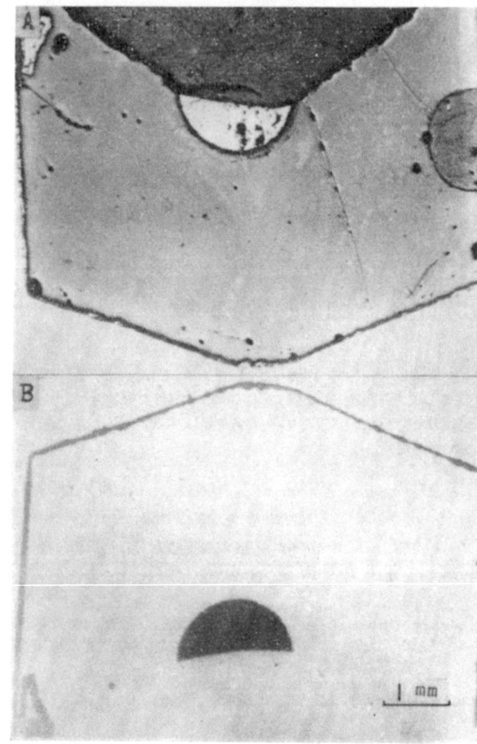

Fig. 2. Photograph (A) and autoradiograph (B) pyrrhotite (fair) and silicate

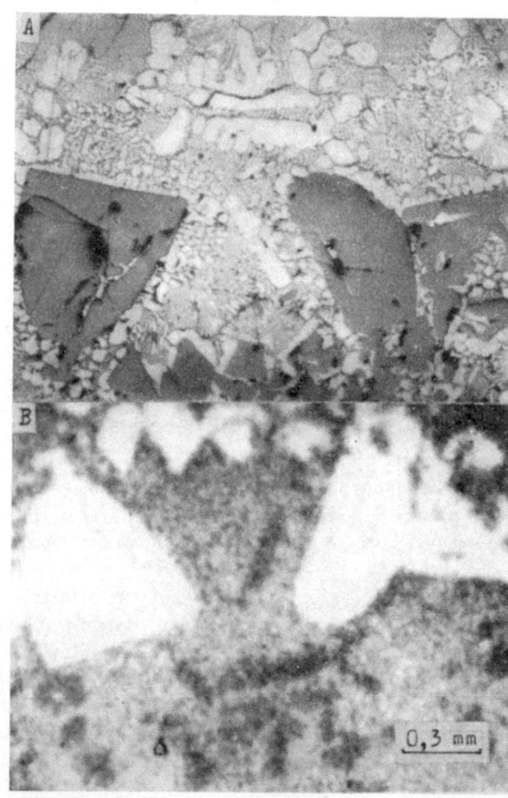

Fig. 3. Photograph (A) and autoradiograph (B) pyrrhotite (fair) and magnetite

The obtained calibration chart allows to determine the partition coefficient of nickel in different phases without recourse to the establishment of its absolute concentrations. The interval of the values, on which the chart is built, makes it possible to find the values of the partition coefficient exceeding three orders while the limiting values do not exceed one-two hundreds during photometry and the optical calculation of the tracks.

The distribution of nickel between the silicate and sulphide, sulphide and metallic phases of the basaltic system as well as between pyrrhotite and magnetite, synthetized from the oxide-sulphide melt can be considered as an example of the use of the instrumental methods.

The experiments on the study of Ni distribution between the phases of the basaltic system were conducted in the resistance furnace with temperature regulators in interval of 1300–1900 °C. 15% of pyrrhotite and the *Clarke* quantities of Ni with the introduced radioisotope ^{63}Ni were added in the charge of the tholeitic basalte. The basaltic glass, pyrrhotite, native iron were obtained after melting and crystallization and crystals of clinopyroxene were obtained only in some experiments. The autoradiographical study showed that Ni is concentrated mostly in iron and less so in pyrrhotite. Microphotometry allowed us to quantitatively determine only one partition coefficient between the metal and pyrrhotite phases is equal 20. Only the lower limit could be indicated for the others. The determination of the quantities of silver in the radiogram substantiated the value of $K_{\text{Fe-Sul}} = 20$, and allowed to establish rather exactly the values of other coefficient, in particular $K_{\text{Sul-Sil}} = 400$ (Fig. 2).

Data on the distribution of Ni between the pyrrhotite and magnetite phases are interesting and new. According to the natural samples and experimental data, obtained with the help of traditional analytical methods, $K_{\text{Pir-Mag}}$ did not exceed 10–15. But as shown by the results of the autoradiographical study, these values are very much underated, evidently, in the main, because of a frequent thin interpretation of these phases and difficulty of their separation pure minerals (Fig. 3). According to the determination of silver in autoradiograms we established the average value of this partition coefficient equal to 25. American scientists obtained similar values also with the help of ^{63}Ni-autoradiography.

Thus, the studies conducted by us show the possibility of silver determination in autoradiograms of home-made plates for the nuclear investigation of MR and MA; the above calibration chart as opposed to the American one, which establishes the dependence of the contents of silver on the exposure value, allows to determine partition coefficient in the investigation phases immediately by silver quantity (in per cent or impulses). The method opens essentially new possibilities of quantitative interpretation of the obtained radiograms, used in the different fields of science and technology.

Received October 18, 1985

References

[1] *T. M. Benjamin, N. T. Arndl, I. R. Holloway,* Instrumental techniques for β-track mapping, Ann. Rept. Dir. Geophys. Lab. Carnegie Inst., 1976–1977, Washington, D.C., 1977, 648–660

Isotopenpraxis 22 (1986) 10, S. 353 bis 356

Synthese und Analyse von [$^{14}CH_3$]Bis(dimethylphosphino)-ethan und dessen Derivaten

P. Mäding, D. Klötzer, R. Münze

(Akademie der Wissenschaften der DDR, Zentralinstitut für Kernforschung, Rossendorf, Bereich Radioaktive Isotope)[1])

Es wird eine Methode zur Synthese von [$^{14}CH_3$]Bis(dimethylphosphino)-ethan beschrieben, die von Ethan-1,2-bis[tris-(hydroxymethyl)-phosphoniumchlorid] ausgeht. Durch stufenweise Abspaltung von Hydroxymethylgruppen und nachfolgender Phosphoniumsalzbildung mit Hilfe von [^{14}C]Methyliodid wird nach mehreren Sequenzen das Syntheseziel erreicht. Für die Radio-DC-Analyse wird ein geeignetes chromatographisches System zur Untersuchung von [$^{14}CH_3$]-Bis(dimethylphosphino)-ethandihydrochlorid und dessen Oxidationsprodukten vorgestellt.

A synthetic method for [$^{14}CH_3$]bis(dimethylphosphino)-ethane is given starting from the ethane-1,2-bis[tris(hydroxymethyl)-phosphoniumchloride]. The synthetic goal is reached by a step-by-step separation of hydroxymethyl groups followed by the formation of the phosphonium salt with the aid of [^{14}C] methyliodide after several sequences. An appropriate chromatographic system for the radio-TC-analysis of [^{14}C]-bis(dimethylphosphino)-ethane dihydrochloride and its oxidation products is presented.

Keywords

carbon 14; labelling; oxidation; phosphines; radiochromatography; synthesis; thin-layer chromatography

1,2-Bis(dimethylphosphino)-ethan (Abk.: DMPE) erwies sich als günstiger Ligand für myocardaffine kationische ^{99m}Tc-Komplexe [1, 2]. An ihrer Nutzung als Radiopharmaka für die Myocard-Perfusions-Szintigraphie wird gegenwärtig gearbeitet [3, 4, 5].

Diese Arbeiten erfordern u. a. pharmakologische Untersuchungen zum Bioverhalten des freien Liganden und seiner Metaboliten. Die Synthese von [$^{14}CH_3$]DMPE und der

[1]) Anschrift: DDR-8051 Dresden, PSF 19

adäquaten radioaktiven Oxidationsprodukte dient diesem Zweck.

Für die ^{14}C-Markierung wurde eine Methode gewählt, die von Ethan-1,2-bis[tris(hydroxymethyl)-phosphoniumchlorid] ausgeht. Diese Verbindung wurde überraschend bei der Salzbildung von 1,2-Bis[bis(hydroxymethyl)-phosphino]-ethan [6] mit alkoholischer HCl in 32%iger Ausbeute erhalten. Eine günstigere Darstellungsweise für das genannte Phosphoniumsalz ist die Umsetzung von 1 Mol 1,2-Bis(phosphino)-ethan mit 6 Mol Formaldehyd und 2 Mol HCl (Ausbeute: 85%).

Zur Charakterisierung von Ethan-1,2-bis[tris(hydroxymethyl)-phosphoniumchlorid] seien folgende Daten angeführt:

Fp: 164 – 168 °C

Elementaranalyse:	$C_{ber.}/C_{gef.}$ = 27·67 %/27·78 %
	$H_{ber.}/H_{gef.}$ = 6.34%/6·77 %
	$Cl_{ber.}/Cl_{gef.}$ = 20·46 %/19·60 %
^{1}H-NMR:	90 MHz, DMSO-d$_6$, TMS ≙ 0, δ in ppm
	2,84 (d, 4H, $-CH_2-CH_2$); 4,04 (s, 6H, OH)
	4,56 (s, 12H, CH_2-O)
^{31}P-NMR:	36,44 MHz, DMSO, H_3PO_4 ≙ 0
	δ = 29,25 ppm (s)
{^{31}P}-^{13}C-NMR:	22,63 MHz, DMSO-d$_6$, TMS ≙ 0, ohne NOE
	7,87 (s, 2C, $-CH_2-CH_2-$);
	50,0 (s, 6C, $-CH_2OH$)
IR(KBr):	ν(OH) = 3600 – 3000 cm^{-1} (st)

Die bevorzugte Abspaltung einer Hydroxymethylgruppe aus Hydroxymethylphosphoniumsalzen durch äquivalente Mengen Alkali gestattet den sukzessiven Aufbau symmetrischer und asymmetrischer tertiärer Phosphine [7]. Infolgedessen wurde der Syntheseweg für die ^{14}C-Markierung gemäß Gl.(1)–(5) konzipiert:

$$[(HOCH_2)_3P{-}C_2H_4{-}P(CH_2OH)_3]\ 2\ Cl \xrightarrow[-2NaCl,2CH_2O]{+2NaOMe} (HOCH_2)_2P{-}C_2H_4{-}P(HOCH_2)_2 \quad (1)$$

$$\xrightarrow{2^xCH_3I} [{}^xCH_3(HOCH_2)_2P{-}C_2H_4{-}P(CH_2OH)_2{}^xCH_3]\ 2I \quad (2)$$

$$\xrightarrow[-2NaI,\ 2CH_2O]{+2NaOMe} {}^xCH_3(HOCH_2)P{-}C_2H_4{-}P(CH_2OH){}^xCH_3 \quad (3$$

$$\xrightarrow{+2^xCH_3I} [({}^xCH_3)_2(HOCH_2)P{-}C_2H_4{-}P(CH_2OH)\ ({}^xCH_3)_2]\ 2I \quad (4)$$

$$\xrightarrow[-2NaI,\ 2CH_2O]{+2NaOMe} ({}^xCH_3)_2P{-}C_2H_4{-}P({}^xCH_3)_2 \quad (5)$$

Die Umsetzungen erfolgten in absolutem Methanol (O_2-frei) unter Argon bzw. Vakuum, wobei die Zwischenprodukte nicht isoliert wurden. Die gewählte spezifische Radioaktivität des eingesetzten [^{14}C]Methyliodid betrug 71 MBq × mmol^{-1}. Das noch verunreinigte [^{14}CH$_3$]DMPE wurde mit alkohol. HCl in sein Dihydrochlorid überführt und an einer mit Al_2O_3 sauer (Fa. Woelm) gefüllten Chromatographiesäule gereinigt. Als Eluationsmittel wurde ein Gemisch aus Methanol/Eisessig/Wasser im Verhältnis 70/30/30 verwendet. Radiochemische Reinheit nach der Säulenchromatographie: 93,6% (Radio-DC).

Nach der Lyophilisierung wurde das gereinigte [^{14}CH$_3$] DMPE · 2 HCl mit Natriummethanolat versetzt und auf diese Weise unter Argon [^{14}CH$_3$]DMPE freigesetzt. Nach Abtrennung des Methanols (im Vakuum bei –40 °C) wurde [^{14}CH$_3$]DMPE zwecks Entfernung einer weiteren mit DMPE nicht mischbaren Verunreinigung an der Vakuumapparatur [8] über P_2O_5/Bimsstein manipuliert, wobei das radioaktive Phosphin mit einer chemischen Reinheit von 97,2% (GC) isoliert werden konnte.

Das gereinigte [^{14}CH$_3$]DMPE wurde erneut mit alkohol. HCl in das Dihydrochlorid umgewandelt (Lyophilisation) und unter Argon gelagert.

Radiochemische Ausbeute: 17%

spezifische Radioaktivität: 270,1 MBq · mmol^{-1}.

Von Bedeutung für die Pharmakologie des DMPE bzw. DMPE · 2 HCl sind auch dessen Oxidationsprodukte, da sie sowohl bei der Präparation der ^{99m}Tc-DMPE-Komplexe, als auch im Organismus aus überschüssigem DMPE entstehen.

Synthese von [^{14}CH$_3$]DMPE-dioxid

Man gewinnt diese Verbindung aus [^{14}CH$_3$]DMPE · 2 HCl durch Oxidation mit verdünntem H_2O_2 im alkalischen Medium.

$$[({}^xCH_3)_2\overset{H}{\overset{|}{P}}{-}C_2H_4{-}\overset{H}{\overset{|}{P}}({}^xCH_3)_2]^{2+}_{2Cl^-} \xrightarrow[-2NaCl,\ 4H_2O]{+2NaOH,\ 2H_2O_2} ({}^xCH_3)_2\overset{O}{\overset{\|}{P}}{-}C_2H_4{-}\overset{O}{\overset{\|}{P}}({}^xCH_3)_2 \quad (6)$$

Aus dem an der Vakuumapparatur getrockneten Reaktionsprodukt wird [^{14}CH$_3$]DMPE-dioxid durch Extraktion mit Chloroform isoliert.

Radiochemische Ausbeute: 73%

spezifische Radioaktivität: 270,1 MBq · mmol^{-1}

Fp: 228–230,5 °C (vorher Sublimation); Lit.-Fp: 232 bis 233 °C [11]

Radio-DC-Analyse von [^{14}CH$_3$]DMPE-dioxid:

Chromatogr. System: Lucefol-Quick; n-Butylacetat/Eisessig/Wasser = 60/40/16

Rf = 0,75

Radiochemische Reinheit: > 98%

Mit diesem DC-System ist es nicht möglich, [^{14}CH$_3$] – DMPE · 2 HCl zu untersuchen.

Oxidation von [^{14}CH$_3$]DMPE · 2 HCl an der Luft

Bei der Lagerung von [^{14}CH$_3$]DMPE · 2 HCl an der Luft entstehen zwei Oxidationsprodukte, die als [^{14}CH$_3$]DMPE-monoxid · 2 HCl und [^{14}CH$_3$]DMPE-dioxid · 2 HCl identifiziert wurden. Die Oxidation verläuft nach folgendem Schema:

$$\underset{(I)}{[{>}\overset{H}{\overset{|}{P}}{-}C_2H_4{-}\overset{H}{\overset{|}{P}}{<}]^{2+}2\,Cl^-} \xrightarrow{[O]} \underset{(II)}{[{>}\overset{HO\cdots\cdots\cdots H}{\overset{|}{P}{-}C_2H_4{-}\overset{|}{P}}{<}]^{2+}2\,Cl^-} \xrightarrow{[O]} \underset{(III)}{[{>}\overset{HO}{\overset{|}{P}}{-}C_2H_4{-}\overset{HO}{\overset{|}{P}}{<}]^{2+}\ 2\,Cl^-} \quad (7)$$

Verbindung II wird bevorzugt gebildet, vermutlich auf Grund von Stabilisierung durch Sechsringbildung über eine intramolekulare Wasserstoffbrückenbindung. Die Identifizierung der Oxidationsprodukte erfolgte u. a. mit Hilfe von inaktiven Vergleichsprodukten.

DMPE-monoxid · 2 HCl wurde durch Oxidation von DMPE · 2 HCl mit verdünntem H_2O_2 erhalten.

Radio-DC-Analyse von [$^{14}CH_3$]DMPE · 2 HCl

Für die Radiodünnschichtchromatographievon [$^{14}CH_3$]-DMPE · 2 HCl und dessen Oxidationsprodukten wurde ein brauchbares chromatographisches System gefunden (Tab. 1), das aus MN-Kieselgel N-HR (Fa. Macherey Nagel u. Co., Düren) und dem Laufmittelgemisch Aceton/Wasser/Methylethylketon/HCl (25%ig) im Verhältnis 60/40/20/1 besteht.

Tab. 1. Radio-DC-Analyse von [$^{14}CH_3$]DMPE · 2 HCl. Chromatographisches System: MN – Kieselgel N-HR; Aceton/Wasser/Methylethylketon/HCl (25 %ig) 60 40 20 1

Substanz	R_f
[$^{14}CH_3$]DMPE-dioxid · 2 HCl	0·3 – 0·35
[$^{14}CH_3$]DMPE-monoxid · 2 HCl	0·45 – 0·5
[$^{14}CH_3$]DMPE · 2 HCl	0·62 – 0·7

Experimentelles

Synthesebedingungen zu den Reaktionsgleichungen (1) – (6)

Gleichung	Ansatz	Reaktionsbedingungen
(1)	0,098 g Na in 4 ml Methanol zu NaOMe umsetzen und 0,738 g Ethan-1,2-bis[tris-(hydroxymethyl)phosphonium-chlorid] zugeben	unter Argon, 1 h rühren, Zimmertemperatur
(2)	604 mg [^{14}C]CH_3I spez. Akt. = 71 MBq · mmol^{-1}	an der Vakuumapparatur [^{14}C]CH_3I auffrieren, unter Vakuum bei 60 – 70 °C 3 1/2 h rühren
(3)	NaOMe aus 0,098 g Na und 3 ml Methanol	unter Argon, Rühren und Eiskühlung zutropfen, unter Vakuum 45 min bei Zimmertemp. und 30 min bei 60 – 70 °C rühren
(4)	592 mg [^{14}C]CH_3I spez. Akt. = 71 MBq · mmol^{-1}	an der Vakuumapparatur [^{14}C]CH_3I auffrieren, unter Vakuum bei 60 bis 70 °C 3 1/2 h rühren
)5)	NaOMe aus 0,102 g Na und 3 ml Methanol	unter Argon, Rühren und Trockeneis/Aceton-Kühlung zutropfen, unter Vakuum bei Zimmertemp. 1 h rühren, danach alle flüchtigen Bestandteile an der Vakuumapparatur isolieren (verunreinigtes [$^{14}CH_3$]DMPE + Methanol)
(6)	0,034 g [$^{14}CH_3$]DMPE · 2 HCl in 1 ml H_2O + 0,35 ml 1 n NaOH + 0,51 ml 3 %iges H_2O_2	45 min unter Rückfluß kochen, Neutralisation mit verd. HCl auf pH = 7. Lösungsmittel an der Vakuumapparatur abziehen, Rückstand mit $CHCl_3$ extrahieren, filtrieren, Lösungsmittel abziehen

Die Anfärbung der nichtradioaktiven Vergleichssubstanzen gelingt mit *Dragendorffs* Reagenz nach *Munier* und *Macheboeff* Nr. 89 [9] und nachfolgender Iodbedampfung. Um eine Oxidation des Phosphoniumsalzes (pk_1 = 5,1 und pk_2 = 7,9 [10]) auf der DC-Platte während des Chromatographieprozesses zu verhindern, wurde im sauren Medium gearbeitet.

Mit Halogenwasserstoffen bilden tertiäre Phosphinoxide die entsprechenden Hydrohalogenide [7].

Mit Hilfe der Radio-DC-Analyse konnte die Stabilität von [$^{14}CH_3$]DMPE · 2 HCl bei unterschiedlichen Lagerbedingungen untersucht werden (Tab. 2).

Gaschromatographische Bedingungen zur Untersuchung des gereinigten [$^{14}CH_3$]DMPE [10]

Gerät: Perkin-Elmer Mod. 900
Säule: 1,8 m, 1/4″-Glassäule
4% Carbowax 20 M + 0,8% KOH
Chromosorb G AW-DMCS (100 — 120 mesh)
Trägergas: H_2, 45 ml/min
Detektor: WLD
Säulentemp.: 120 °C

Zusammenfassung

Die Synthese von [$^{14}CH_3$]DMPE gelingt aus Ethan-1,2-bis[tris(hydroxymethyl)-phosphoniumchlorid] durch stufenweise Abspaltung der Hydroxymethylgruppen mit Natriummethanolat und jeweils anschließender Phosphoniumsalzbildung mit [^{14}C]Methyliodid. Während bei niederen und mittleren spezifischen Radioaktivitäten des [^{14}C]Methyliodids (z.B. 70 MBq · mmol^{-1}) brauchbare Syntheseergebnisse erzielt werden, beeinträchtigen bei hohen spezifischen Aktivitäten des [^{14}C]Methyliodids (z. B. 400 MBq · mmol^{-1}) Radiolyseprozesse den Syntheseverlauf enorm.

Das Syntheseprodukt muß unter Argon gelagert werden, da es im beträchtlichen Maße der Oxidation unterliegt. Bei der Oxidation des Phosphoniumsalzes an der Luft entsteht zunächst [$^{14}C_3$]HDMPE-monoxid · 2 HCl als relativ stabile Zwischenstufe. [$^{14}CH_3$]DMPE-dioxid gewinnt man aus [$^{14}CH_3$]DMPE·2 HCl durch Oxidation mit verdünntem H_2O_2 im alkalischen Milieu.

Ein für die Radiodünnschichtchromatographie dieser Verbindungen geeignetes System wird beschrieben.

Eingegangen am 18. 12. 1985

Literatur

[1] *E. Deutsch, K. A. Glavan, V. J. Sodd, H. Nishiyama, D. L. Ferguson, S. J. Lukes,* J. Nucl. Med. 22 (1981) 897

[2] *E. Deutsch, K. Libson, J. L. Vanderheyden, D. L. Nosco, V. J. Sodd, H. Nishiyama,* J. Nucl. Med. 23 (1982) 9

Tab. 2. Analyse von [$^{14}CH_3$]DMPE · 2 HCl nach Lagerung unter unterschiedlichen Bedingungen

		Radio-DC-Analyse		
Lagerzeit	Lagerbedingungen	[$^{14}CH_3$]DMPE · 2 HCl	[$^{14}CH_3$]DMPE-monoxid · 2 HCl	[$^{14}CH_3$]DMPE-dioxid · 2 HCl
6 Monate	Argon	97 %	3·0 %	–
23 Stunden	Luft	93·5 %	6·5 %	–
3 Monate	Luft	15·0 %	65·0 %	20·0 %

[3] *M. C. Gerson, E. A. Deutsch, H. Nishiyama, K. F. Libson, R. J. Adolph, G. W. Grossman, V. J. Sodd, D. L. Forstmann, J. L. Vanderheyden, C. C. William, E. L. Saenger,* Eur. J. Nucl. Med. 8 (1983) 371

[4] *R. Syhre, S. Seifert, R. Münze, D. Klötzer, P. Mäding, R. Berger, K.-H. Heise, R. Jander, M. Kretzschmar,* Nuc. Compact **15** (1984) 130

[5] *R. Syhre, S. Seifert, R. Münze, R. Jander, M. Kretschmar,* Nuc. Compact **15** (1984) 240

[6] *D. Klötzer, P. Mäding, R. Münze,* Z. Chem. **24** (1984) 224

[7] *Houben-Weyl-Müller,* Methoden d. org. Chemie. Bd. XII/1, Stuttgart: Verlag Georg Thieme 1963

[8] *M. Bubner, L. Schmidt,* Die Synthese Kohlenstoff-14-markierter organischer Verbindungen, Leipzig: VEB Georg Thieme 1966

[9] *E. Stahl*: Dünnschichtchromatographie. Berlin, Heidelberg, New York: Springer-Verlag 1967

[10] *S. Seifert, D. Klötzer, P. Mäding, R. Münze, K.-H. Heise, H. Görner,* Jahresbericht Zentralinstitut für Kernforschung, Rossendorf, *R. Münze* (ed.) ZfK-570 (im Druck)

[11] *G. M. Kosolapoff, R. F. Struck,* J. Chem. Soc. (1961) 2423

Isotopenpraxis 22 (1986) 10, S. 356 bis 361

Zur Bestimmung der Schichtdicke von Flüssigkeitsfilmen unter Anwendung offener Radionuklide

J. Stopporka[1], *F. Häußler*[2], *E.-J. Langrock*[2]

[1] Technische Hochschule Leipzig, Sektion Polygrafie
[2] Technische Hochschule Leipzig, Sektion Naturwissenschaften

Zur dynamischen Schichtdickenbestimmung dünner und vornehmlich flüssiger Schichten existieren für viele Anwendungsfälle keine technisch nutzbaren Lösungen. Im vorliegenden Artikel wird deshalb die Anwendbarkeit offener, γ-strahlender Radionuklide für eine nichtisotope, physikalische Markierung untersucht. Bei diesen Tracern ist ein Zusammenhang zwischen der Schichtdicke und der ermittelten Zählrate zu erkennen. Mit ^{99m}*Tc ist die Meßaufgabe unter günstigen geometrischen Bedingungen lösbar. Für allgemeine Fälle werden Radionuklide mit höheren γ-Energien empfohlen. Außerdem wird theoretisch und praktisch eine Optimierung des Meßaufbaus an einer rotierenden Walze vorgenommen.*

For dynamic thickness measurements of thin and particularly liquid films, in many cases, there exist no technical useful solutions. For this reason, the applicability of open γ-radiating radionuclides to a non-isotopic physical labelling is investigated in the paper. With these tracers, a correlation between the film thickness and the measured counting rate can be found. The measuring-problem can be readily solved with ^{99m}*Tc under favourable geometric conditions. For general cases, radionuclides with higher γ-energies are recommended. Moreover, a theoretical and practical optimization of a measuring set-up at a rotating roll is carried out.*

Keywords

industry; measuring methods; radioisotope generators; technetium 99m; thickness; tracer techniques

1. Einleitung

In vielen Bereichen der Technik ist die Bestimmung der Dicke dünner Schichten zur Untersuchung von Prozeßabläufen oder zur Kontrolle vorgegebener Parameter erforderlich.

Für einen speziellen Anwendungsfall wurde ein dynamisch arbeitendes Verfahren gesucht, das die Bestimmung der Dicke von flüssigen Schichten ermöglicht. Darüber hinaus wurden folgende Forderungen gestellt:

- Messung auf bewegtem, planem oder unebenem Trägermaterial bei verschiedensten Geschwindigkeiten
- Messung auf Unterlagen verschiedener Rauhigkeit und unterschiedlicher Materialzusammensetzung
- Meßempfindlichkeit von etwa 1 µm bei einem Meßbereich bis zu 60 µm
- Messung von transparenten, elektrisch leitfähigen Flüssigkeitsschichten
- Verfahren ohne bzw. mit vernachlässigbar kleinem Einfluß, der sich über der Schicht ändernden Luftfeuchtigkeit
- Verfahren ohne Beeinflussung des Schichtbildungsprozesses bzw. des Schichtmaterials oder der Schichtdicke.

Anschrift: DDR-7030 Leipzig, Karl-Liebknecht-Str. 132

Tab. 1. Für technische Untersuchungen geeignete Radionuklidgeneratorsysteme

Mutternuklid		Tochternuklid		γ-Energien in keV in ( ) Intensität in %
Nuklid	Halbwertszeit	Nuklid	Halbwertszeit	
^{137}Cs	30,14 a	^{137m}Ba	2,55 min	661,6 (89,9)
^{68}Ge	270,8 d	^{68}Ga	68,1 min	1077,4 (2,93)
^{113}Sn	115,1 d	^{113m}In	1,66 h	391,7 (64,2)
^{87}Y	3,35 d	^{87m}Sr	2,8 h	388,4 (81,8)
^{44}Ti	47,3 a	^{44}Sc	3,93 h	1157,0 (99,9)
^{169}Yb	2,36 d	^{169}Tm	7,7 h	2052,4 (20,5)
				1273,4 (17,4)
				1176,6 (10,9)
^{100}Pd	3,63 d	^{100}Rh	20,8 h	2376,1 (35,0)
				1929,7 (12,2)
				1553,4 (20,5)
				1362,1 (15,0)
				1107,1 (13,2)
^{194}Hg	520 a	^{194}Au	39,5 h	328,4 (63,0)
^{140}Ba	12,75 d	^{140}La	40,28 h	1596,5 (95,4)
^{172}Hf	1,87 a	^{172}Lu	6,7 d	1093,6 (68,9)
				912,1 (16,8)
				900,7 (32,9)

Auf Grund nicht verfügbarer anderer Meßprinzipien wurden hier nur die Einsatzmöglichkeiten von Methoden unter Nutzung der Radioaktivität untersucht.

Mitteilungen / Information

IAEA: BEVORSTEHENDE IAEA-TAGUNGEN UEBER KERNSICHERHEIT

Die IAEA gibt folgende Termine fuer eine Reihe von Sondertagungen ueber Kernsicherheit als Reaktion auf den Unfall in Tschernobyl bekannt:

1) Tagung zur Ausarbeitung von Entwuerfen internationaler Abkommen,
21.Juli bis 8. August 1986, im Amtssitz der IAEA in Wien. Bei dieser Tagung von Regierungssachverstaendigen werden die Entwuerfe fuer zwei internationale Abkommen ausgearbeitet werden. Durch das eine Abkommen wuerden die Vertragsparteien dazu verpflichtet, bei nuklearen Unfaellen mit moeglichen grenzueberschreitenden Auswirkungen Fruehwarnung und Informationen zu geben; in diesem Abkommen wuerden die derzeit geltenden, in der Informationsschrift INFCIRC/321 festgelegten Richtlinien der IAEA beruecksichtigt. Durch das andere Abkommen waeren die Parteien verpflichtet, Noteinsaetze und Hilfsmassnahmen bei nuklearen Unfaellen zu koordinieren; in diesem Abkommen wuerden die in der Informationsschrift INFCIRC/310 festgelegten Richtlinien beruecksichtigt.

2) Eroerterung des Unfalls,
25.-29. August 1986, im Amtssitz der IAEA in Wien. Bei dieser Tagung, an der Sachverstaendige aus der Sowjetunion und anderer Mitgliedsstaaten sowie Vertreter internationaler und zwischenstaatlicher organisationen teilnehmen werden, sollen die ueber den Unfall von Tschernobyl zur Verfuegung stehenden Informationen eroertert werden. Die Ergebnisse und Empfehlungen der Tagung werden dem Gouverneursrat der IAEA noch vor seiner ordentlichen Tagung vom 22.-23. September 1986 vorgelegt werden.

3) Ausserordentliche Tagung der IAEA-Generalkonferenz,
24.-26. September 1986, im Kongresszentrum der Hofburg in Wien. Diese ausserordentliche Tagung wird noch vor der 30. ordentlichen Tagung der IAEA-Generalkonferenz stattfinden, die am 29. September, ebenfalls in der Wiener Hofburg, eroeffnet wird und bis 3. Oktober dauern soll. Im Rahmen dieser ausserordentlichen Tagung werden Massnahmen fuer eine staerkere internationale Zusammenarbeit auf dem Gebiet der Kernsicherheit und des Strahlenschutzes besprochen werden.

4) Arbeitsgruppe von Sachverstaendigen fuer Kernsicherheit,
wird voraussichtlich im November 1986 stattfinden, der genaue Termin wird noch bekanntgegeben. Dies wird die erste Tagung einer Arbeitsgruppe von Sachverstaendigen sein, die, ueber einen laengeren Zeitraum hinweg zusaetzliche Massnahmen zur Verbesserung der Zusammenarbeit auf dem Gebiet der Kernsicherheit, darunter auch Mittel und Wege zur Verfeinerung der nuklearen Sicherheitsnormen, eroertern soll.

Um das ausgeweitete Kernsicherheitsprogramm der IAEA abzudecken, hat der IAEA-Gouverneursrat US$ 679.000 bis Ende 1986 und bis zu US$ 2 Millionen, fuer 1987 genehmigt.

Diese Betraege kommen zu den etwa US$ 6 Millionen, die bereits vorher im ordentlichen Budget der IAEA fuer 1987 fuer Kernsicherheitsaktivitaeten vorgesehen waren, noch hinzu. So wie es vom Gouverneursrat der IAEA-genehmigt wurde, belaeuft sich das ordentliche Gesamtbudget der IAEA fuer 1987 (fuer alle Taetigkeitsbereiche) auf knapp unter US$ 104 Millionen. Gemaess der Satzung der IAEA wird das Budget nun der IAEA-Generalkonferenz zur endgueltigen Genehmigung vorgelegt werden. (IAEA-Pressekommunique (1986)12)

IAEA: DURCHFUEHRUNG DER SICHERUNGSMASSNAHMEN IM JAHRE 1985

Das Sekretariat der IAEA hat im Laufe des Jahres 1985 keine Anomalie entdeckt, "die auf die Abzweigung einer signifikanten Menge von den IAEA-Sicherungsmassnahmen unterstehendem spaltbarem Material oder auf den Missbrauch von den Sicherungsmassnahmen unterstehenden Anlagen oder Geraeten zum Zwecke der Herstellung einer Kernwaffe oder zu irgendeinem anderen militaerischen Zweck oder zur Herstellung irgendeines anderen nuklearen Sprengkoerpers oder zu unbekannten Zwecken hingedeutet haette."

Diese wesentliche Schlussfolgerung wird im jaehrlichen Bericht des IAEA-Sekretariats ueber die Durchfuehrung der Sicherungsmassnahmen (Safeguards Implementation Report) an den Gouverneursrat gezogen, der den Bericht in seiner Junitagung in Wien behandelt hat. Das Ziel der IAEA-Sicherungsmassnahmen ist es, sicherzustellen, dass den Sicherungsmassnahmen unterstehende nukleare Programme und Nuklearanlagen nicht fuer militaerische Zwecke missbraucht werden.

Die Genauigkeit der Inspektions- und Auswertungstaetigkeit kann durch die Tatsache veranschaulicht werden, dass 1985 ungefaehr 150 (400 im Jahr 1984) groesstenteils kleinere Diskrepanzen und Anomalien festgestellt wurden. Alle Faelle konnten bei anschliessender Ueberpruefung oder Nachforschung zufriedenstellend erklaert werden. 1985 wurden im Rahmen von Sicherungsabkommen mehr als 1980 Inspektionen in 514 Kernanlagen in 51 Nichtkernwaffenstaaten und vier Kernwaffenstaaten durchgefuehrt. (Die Zahlen fuer 1984 lauten: 1820 Inspektionen in 474 Anlagen in 49 Nichtkernwaffenstaaten und drei Kernwaffenstaaten.) Bei 32% (30% im Jahr 1984) der Inspektionen wurde das Kernmaterial mittels zerstoerungsfreier Messverfahren geprueft. Mehr als 290 (240) automatische Photo- und Fernsehueberwachungssysteme waren in Nuklearanlagen im Einsatz, und 9000 (8300) zum Verschluss von Kernmaterial angebrachte Siegel wurden abgenommen und anschliessend im Amtssitz der IAEA auf ihre Unversehrtheit ueberprueft. Mehr als 1270 (1080) Pu- und U-Proben wurden analysiert, und davon wurden ungefaehr 3250 (2770) Analyseergebnisse mitgeteilt. Im IAEA-Computer wurden 561.000 (760.000) Eintragungen von Duchfuehrungs- und anderen Sicherungsdaten verarbeitet und gespeichert. Am Ende des Jahres 1985 belief sich die Menge des den IAEA-Sicherungsmassnahmen unterstehenden Kernmaterials, inklusive jenem, das durch Abkommen mit Kernwaffenstaaten erfasst ist, auf 9 Tonnen (7,7t) abgetrenntes Pu; 155,1t (129,5t) in bestrahlten Brennstoffen enthaltenes Pu; 12,3t (11,8t) hochangereichertes U; 24.546t (22.784t) niedrigangereichertes U und 43.044t (31.724t) Ausgangsmaterial, darunter auch abgereichertes U und Th. Ende 1985 waren insgesamt 163 Sicherungsabkommen mit 96 Staaten in Kraft, im Vergleich zu 163 Abkommen mit 95 Staaten Ende 1984. In Nichtkernwaffenstaaten unterstanden 474 Kernanlagen den Sicherungsmassnahmen oder enthielten Material, das Sicherungsmassnahmen unterstand. (IAEA-Pressekommunique (1986) 13)

IAEA: GELDMITTEL FUER DIE TIERFORSCHUNG

Die Regierung der Niederlande hat beschlossen, ein FAO/IAEA-Forschungsprogramm mit ungefaehr US$ 1 Million ueber eine Anfangsperiode von drei Jahren zu foerdern. Im Rahmen dieses Forschungsprogramms werden Veterinaermediziner und Tierforscher in Afrika Ausbildungsmoeglichkeiten und Einrichtungen zur Verfuegung stehen, um die erforderliche Feldforschung und Laborarbeit zum Nachweis des Status hinsichtlich durch das Blut uebertragener Krankheiten und hinsichtlich des Fortpflanzungszyklus bei Wiederkaeuern in kleinen Landwirtschaftsbetrieben durchzufuehren. Nach Abschluss dieses Forschungsprogramms wird es den Wissenschaftlern moeglich sein, Aenderungen in der Tierhaltung auszuprobieren und auch zu empfehlen, wenn die afrikanischen Kleinzuechter dadurch das Potential ihres Viehbestandes besser ausschoepfen koennen.

In gleicher Weise wird die Schwedische Internationale Entwicklungsbehoerde (SIDA) nach endgueltiger Genehmigung durch die schwedische Regierung mit ungefaehr US$ 600.000 ueber einen Zeitraum von fuenf Jahren ein internationales Forschungsprojekt im Rahmen des Gemeinsamen FAO/IAEA-Programms unterstuetzen. Durch dieses Projekt sollen Immunoassaymethoden zur Diagnose von Virus- und Bakterieninfektionen wie zum Beispiel Rinderpest und Brucellose entwickelt werden. Diese Methoden koennen dann zur Beobachtung in Erkrankungsfaellen und als Hilfsmittel fuer Veterinaerbehoerden bei der Bekaempfung von Krankheiten durch Impf- oder Ausrottungsaktionen angewendet werden.
Aktionen dieser Art sind notwendig, weil ein grosser Anteil des weltweiten Bestands an wiederkaeuendem Vieh (z.B. Rinder, Schafe und Ziegen) in Entwicklungslaendern gezuechtet wird, und zwar grossteils von Kleinbauern, die von diesen Tieren abhaengig sind, da sie nicht nur zur Gewinnung von Fleisch, Milch und anderen Produkten wie zum Beispiel Wolle und Leder fuer die Versorgung ihrer Familien dienen, ihnen ein Einkommen verschaffen, mit dem Samen und andere lebensnotwendige Haushaltgueter gekauft werden koennen, sowie als Zugtiere und Duengemittellieferanten wirken. Eine Bedrohung fuer alle Kleinzuechter stellt das Auftreten von Krankheiten dar, die von Blutparasiten, Viren und Bakterien bei ihrem Tierbestand hervorgerufen werden. Selbst wenn diese Krankheiten nicht toedlich wirken, soe setzen sie doch die Faehigkeit der Tiere, den Zuechterhaushalt mit lebensnotwendigen Produkten zu versorgen, betraechtlich herab. Ein anderer Faktor, welcher nicht unbedingt mit einer Krankheit zusammenhaengen muss, der aber schwerwiegende Folgen fuer den Zuechter hat, ist der, dass sich die Tiere zu langsam vermehren. Eine niedrige Fortpflanzungsrate bedeutet natuerlich, dass es weniger Jungtiere gibt, die verkauft oder durch die anderen Tiere ersetzt werden koennen.
Da die Jungtiere auch die Milchproduktion ausloesen, kommt es bei nicht regelmaessiger Fortpflanzung auch zu einer betraechtlichen Verminderung der Milchproduktion. Radio- und Enzymimmunoassaymethoden in Verbindung mit kuerzlich gemachten biotechnologischen Errungenschaften haben zu sehr einfachen und verlaesslichen analytischen Methoden gefuehrt, um Krankheitserreger zu entdecken und festzustellen, ob ein Tier traechtig ist oder nicht. Diese Methoden koennten somit die Informationen liefern, die benoetigt werden, um Krankheiten zu bekaempfen und die Fortpflanzungsfaehigkeit der Tiere zu verbessern, und langfristig gesehen, die Produktivitaet der Tiere anzuheben und dadurch auch die laendliche Entwicklung zu foerdern. (IAEA-Pressekommunique (1986) 11)

Kurznachrichten / Newlets

Leninpreise 1986:
IN DER VORHUT DER WISSENSCHAFT

Die Traeger des Lenin- und des Staatspreises der UdSSR gehoeren zur Vorhut der schoepferischen Kraefte in der Sowjetunion. Eine Arbeit von A. Andrejew und anderen fusst auf der Entdeckung einer neuen Erscheinung in der Physik und ihrer experimentellen Bestaetigung durch die Autoren - des Materialtransports mittels Tunneleffekt und der Quantenkristallisation. Frueher nahm man an, dass sich in Festkoerpern jeder Transport von Stoffen, beispielsweise die Bildung von Keimen einer neuen Phase usw., nur so vollziehen kann, dass die Energie der Waermebewegung der Atome mitunter die Barriere der potentiellen Energie im Stoff ueberschreitet und die Atome auf diese Weise diese Barriere ueberwinden koennen. Diese klassische Diffusionsuebertragung von Stoffen nimmt haeufig bei Temperaturanstieg zu und geht bei Abkuehlung zurueck. Mit diesem Prozess haengen viele Eigenschaften von Stoffen zusammen, so der Uebergang von einer kristallinen Form in eine andere, die Verteilung von Fremdstoffen usw.
Die Autoren entdeckten noch einen vollkommen anderen Prozess, und zwar einen reinen Quantenprozess. Hierbei handelt es sich darum, dass ein Teilchen die Barrieren mit Hilfe des Tunneleffekts, von Durchgaengen unter den Barrieren, ueberwinden kann, wenn die Energie des Atoms geringer ist als die Groesse der Barriere. Dieser Effekt nimmt mit sinkender Temperatur zu (im Gegensatz zur herkoemmlichenDiffusion), er kann zur Quantenkristallisation eines Stoffes fuehren, die sich mit sinkender Temperatur beschleunigt. Daraus ergeben sich zahlreiche neue, unerwartete Eigenschaften des Stoffes. Im Grund genommen ist das ein neues Kapitel in der Festkoerperphysik.
Eine andere Arbeit auf dem Gebiet der Physik von S. Denissow und anderen steht unter dem Thema "Confinement-Prozesse bei starken Wechselwirkungen von Elementarteilchen mit hohen Energien". In diesem Zyklus wird ein neues Herangehen an die Untersuchung der Mehrfachbildung von Hadronen, in starker Wechselwirkung befindlicher Teilchen vorgeschlagen.
In der Chemie ist die hohe Auszeichnung der Arbeit "Syntheseprozesse und Eigenschaften von Polymeren" von W. Korschak zuerkannt worden. Der Autor entwickelte Methoden zur Synthese einiger elementorganischer Polymere wie bororganischer, phosphororganischer und anderer. Die Arbeiten des Autors bilden einen wesentlichen Beitrag zur industriellen Produktion von Polymerwerkstoffen in der UdSSR. Eine bedeutsame Grundlagenarbeit stellt auf dem Gebiet der Chemie die Arbeit "Magnet-Spin-Effekte bei chemischen Reaktionen" von J. Molin und anderen dar. Jahrelang hatte man angenommen, dass Magnetfelder den Verlauf chemischer Reaktionen nich beeinflussen koennen, da die wirkenden Kraefte um vieles kleiner sind als die elektrischen Kraefte, die zwischen den Atomen oder Ionen wirken. Die Entdeckung der Autoren bestand darin, dass sie den Einfluss der Magnetfelder auf die Spins (magnetische Momente) der Elektronen der reagierenden Teilchen feststellten. Diese Arbeiten gelangten zu weltweiter Anerkennung und fuehrten zur Entstehung eines neuen Gebiets in der Chemie. Ebenfalls gewuerdigt wurden in der Chemie die in Gorki von G. Dewjatych ausgefuehrten Forschungsarbeiten zu den Methoden der Gewinnung von hochreinen Stoffen. Auf der Grundlage des neuen Verfahrens erhielt man mehrere Stoffe mit einem Gesamtinhalt von Fremdstoffen von 10^{-4} bis 10^{-5} Prozent. Der Reinheitsgrad einiger Stoffe uebertrifft den bei analogen Stoffen im Ausland erzielten. (Presse der Sowjetunion (1986) 8)

Radiological Protection:
DOSE LIMITS REVISED

New GUIDANCE on dose limits for members of the public has been published by the National Radiological Protection Board. It endorses a statement issued by the International Commission on Radiological Protection (ICRP) in 1985 and considers its practical application. The ICRP statement requires that two dose limits for the public are used. The principal dose limit is the annual dose averaged over a lifetime - 1 mSv. The subsidiary dose limit is that in a single year - 5 mSv. There is difficulty in demonstrating compliance with a limit on annual dose averaged over a lifetime. Hence, while reaffirming the advice given previously, to keep lifetime dose below 70 mSv, the NRPB recommends that, in practice, the simplest way to ensure compliance is to apply a single annual limit on dose of 1 mSv. The dose to be compared with this limit is the sum of the effective dose equivalent received from external radiation in the year and the committed effective dose equivalent from intakes of dose equivalent from intakes of radionuclides in the same year. Use of this limit of 1 mSv in a year means

that the single organ dose limit for members of the public (50 mSv) is unnecessary for those organs or tissues specified in the definition of effective dose equivalent. However, some nuclides will be limited by the dose to the skin or lens of the eye, to which effective dose is not applicable, and the Board recommends continued use of its limits of 50 and 15 mSv, respectively. As a result of this advice the Board will undertake new calculations of generalised derived limits (GDLs), which can be used as simple screening values for levels of radionuclides in the environment. For each radionuclide and environmental materials GDLs will be specified based on the 1 mSv dose or on the annual dose limits for skin or lens of the eye, whichever is the most restrictive. GDLs will continue to be based on the dose to the critical age group, taking into account variations in dose per unit intake, and dietary and other habits with age. The age groups considered continue to be infants (one year old), children (10 years old) and young adults (20 years old) as well as infants on a milk diet.
(Atom (1986) 355)

Veröffentlichungen / Publications

FROM THE CENTRAL RESEARCH INSTITUTE OF PHYSICS, HUNG. ACADEMY SCI., BUDAPEST, HUNGARY

Preprints
KFKI-1986-25/E
Theory of the nonstationary phase conjugation by four-wave mixing - G.P. Djotyan et al

KFKI-1986-26/E
A new approach to the problem of chaotic repellers - P. Szepfalusy, T. Tel

KFKI-1986-27/E
Peak shape determination in laser microprobe mass analysis - A. Vertes et al

KFKI-1986-28/E
Variational ground state for the periodic Anderson model - P. Fazekas

N E U E R S C H E I N U N G E N

Gruppentheorie fuer Chemiker
VCH Verlagsges., Weinheim 1985, 234 S., D. Wald

Handbook on the Physics and Chemistry of the Actinides
Vol. 3, edited by A.J. Freeman, C. Keller - North Holland Physics Publ., Amsterdam 1985, pp. 520

IAEA - A-1400 Vienna, Wagramerstrasse 5, P.O. Box 100, Austria
IAEA Safety Codes and Guides (NUSS) in the Light of Current Safety Issues

Plasma Physics and Controlled Nuclear Fusion Research 1984
Vol. I:
Artsimovich memorial lecture; Tokamak experiments; Plasma heating and current drive
Vol. II:
Magnetic confinement theory; Open confinement systems; Alternative and supporting systems
Vol. III:
Inertial confinement; INTOR; Technology and reactor concepts; Fundamental processes and new trends; Summaries

Veranstaltungskalender – Tagungen – Lehrgänge / Calendar – Meetings – Courses

NATIONALE TAGUNG UEBER DIE ANWENDUNG NUKLEARER METHODEN IN TIERPRODUKTION UND VETERINAERMEDIZIN
vom 26.-27. November 1986 in Leipzig, DDR
Veranstalter:
Wiss. Gesellschaft fuer Veterinaermedizin der DDR und Karl-Marx-Universitaet Leipzig, Sektion Tierproduktion und Veterinaermedizin, Lehrstuhl fuer Radiologie, Zentrales Isotopenlabor
DDR-7010 Leipzig, Semmelweisstrasse 4

INTERNATIONALE KONFERENZ UEBER ENERGIEUMWANDLUNGEN UND WECHSELWIRKUNGEN MIT KLEINEN UND MITTELRAEUMIGEN ATMOSPHAERISCHEN VORGAENGEN
vom 2.-6. Maerz 1987 an der Technischen Hochschule Lausanne, Schweiz (SVA-Bulletin (1986) 10/11)

11th RADIOCHEMICAL CONFERENCE
will be heldt at Mariánské Lázně, CSSR, April 21-25, 1987
The following subjects have been proposed for the presentation:
- Radioanalytical methods
- Radiochemical separation methods
- Preparation of labelled compounds
Futher information: Dr. V. Spěváčková, Department of Nuclear Chemistry, Technical University of Prague, Brehova 7, CS-115 19 Praha 1

Ausgewählte Beiträge / Recently published papers

EUROP. J. NUCL. MED. 11(1985)5
Nephrography with radioactive hippuran in transplanted kidneys : Interpretation, limitations and usefulness - H. Fill et al

KHIMIKO FARMAZEVTITSCHESKII ZHUR. 19(1985)9
Study of the metabolism of oxoplatinum-Pt-191 - A.S. Singin et al

VOPROSI MEDITSINSKOI KHIMII 31(1985)5
Isolation and properties of chymotrypsin and trypsin labeled with iodine radioisotopes - V.N. Kulakov et al

JUGOSLAVICA PHYSIOLOGICA ET PHARMACOLOGICA ACTA 21(1985)2
Effects of Co-60 irradiation and chemotherapentic agent VM 26 on the induction of thermotolerance in isolated mammalian cells - O. Dordevic et al

Incorporation of P-32 into nucleic acids and proteins during compensatory renal growth in mice - H. Banfic et al

INTERN. J. RADIAT. BIOLOGY 48(1985)4
I-131-induced DNA double strand breaks : use in calibration of the neutral filter eluation technique and comparison with X-ray induced breaks - L.R. Radford; G.S. Hodgson

Radiation-induced strand breaks in FX 174 replicative form DNA: an improved experimental and theoretical approach - J.H. van Touw et al

CLINICA CHIMICA ACTA 151(1985)1
Comparison of N-15 analysis by optical emission spectrometry and mass spectrometry for clinical studies during total parentenal nutrition - A. Ragon et al

FARMAKOLOGIYA TOKSIKOLOGIYA 48(1985)5
The effect of radiographic contrast reagents on bilirubin absorption by rat liver sections - E.N. Bolotova et al

SCAND. J. CLIN. LAB. INVEST. 45(1985)6

The double isotope technique for in vivo determination of the tissue-to-blood partition coefficient for xenon in human subcutaneons adipose tissue - an evaluation - R. Jelnes et al

ACTA MEDICA OKAYAMA 39(1985)5
Metabolism of branched-chain amino acids in rats with acute hepatic failure : A tracer study using N-15-leucine - H. Usui et al

AMER. J. BOTANY 72(1985)8
Loss and uptake of N-15-ammonium in submerged soils of a cattail marsh - J.V. Dean et al

XENOBIOTICA 15(1985)10
Peroxidase - catalyzed binding of U-C-14-phenol to DNA - V.V. Subramonyam et al

BIOLOGICAL PSYCHIATRY 20(1985)11
Platelet H-3-imipramine binding in depressed elderly patients - L.S. Schneider et al

J. PEDIATRIC GASTROENTEROL. NUTRITION 3(1984)5
Urea utilization by the intestinal flora, of infants fed mother's milk and a formula diet, as measured with the N-15-tracer technique - W. Heine et al

BJULL. EKSPERIMENT. BIOLOGII MEDIZINY 99(1985)1 (in Russian)
Scope for some enzymic tests and for radionuclide hepatography in the diagnosis of early changes in liver function - N. Vlachov et al

EUROP. J. NUCL. MED. 11(1985)2-3
Nuclear medicine liver-function tests for pregnant women and children. 2. A new test method via urine usine N-15-methacetin - P. Krumbiegel et al

A liver-function test using N-15 labelled ammonium chloride - K. Jung et al

J. BIOL. CHEM. 260(1985)17
N-15-labeled Escherichea coli. Double resonance and two-dimensional NMR of N-15-labeled psendouridine - R.H. Griffey et al

CANCER LETTERS 27(1985)3
Metabolic studies of N-15-labelled N-nitrosoproline in isolated rat hepatocytes and subcellular fractions - S.R. Koepke et al

WISSENSCHAFT UND FORTSCHRITT 36(1986)2
Chemie - eine Quelle unseres Wohlstandes oder "Giftkueche"? - W. Schirmer; L. Ebur

Metall-Polymer-Verbunde - C. Bischof

Kometenforschung - D. Moehlmann

Der quantisierte Hall-Effekt, Physik-Nobelpreis 1985 - R. Herrmann

WISSENSCHAFT UND FORTSCHRITT 36(1986)3
30 Jahre VIK Dubna: Muster fuer Forschungskooperation - S. Tesch

Element 108 und 109, Synthese am Dubnaer Schwerionenzyklotron - H. Bruchertseifer

Direkte Methoden zur Kristallstrukturanalyse, Nobelpreis fuer Chemie 1985 - L. Kutschabsky; K.-O. Backhaus

Forschungsobjekt Antarktis - H. Kautzleben, R. Frey

WISSENSCHAFT IN DER UDSSR (1986) 2
Wie man die Ionenkanaele steuern kann - L. Magasanik

Neutronengas und seine Aufbewahrung - W. Morosow

SPECTRUM 17(1986)2
Energie aus dem Mikrokosmos, 30 Jahre Zentralinstitut fuer Kernforschung (ZfK), Rossendorf - G. Flach

SPECTRUM 17(1986)3
Polyelektrolytforschung - Probleme, Ergebnisse, Anwendungen - B. Philipp

Pruefen, ohne zu zerstoeren - G. Frankowski

SPECTRUM 17(1986)4
Physik und Schluesseltechnologien - G. Albrecht

"Farbige Quarks" und schwere Ionen - Exklusiv-Interview mit N.N. Bogoljubow

Dubna-Wahlheimat und zweiter Arbeitsplatz - K. Hennig

SPECTRUM 17(1986)5
Die Suche nach schweren Elementen - E. Manke

ANGEW. CHEMIE 98(1986)1
Die 2 4-Dreifachbindung zwischen Mo und Wo - eine anorganische funktionelle Gruppe - M.H. Chrisholm

Die Chemie des nicht-kristallinen Zustandes - S.R. Elliott et al

Molekularbiologische Grundlagen der Spezies - und Organspezifitaet von Influenza - A - Viren - C. Scholtissek

Periodische Potentialflaechen in Kristallstrukturen - R. Nesper, H.G. von Schnering

ANGEW. CHEMIE 98(1986)2
Physikochemische und cytochemische Untersuchungen zur Bindung von Ethidium- und Acriolinfarbstoffen an DNA und an Organellen in lebenden Zellen - H.W. Zimmermann

Resonanz-Raman-Spektroskopie (RR-Spektroskopie) und ihre Anwendung in der Anorganischen Chemie - R.J.H. Clark, T.J. Dines

CF_2-verbrueckte Metallkomplexe - W. Schulz et al

Stabile Selenocarbonyl-ylide (neue organische Zeile) - R. Allmann et al

ANGEW. CHEMIE 98(1986)3
Chemie mit Clusterstrahlen - von Atomen zum Festkoerper - T.P. Martin

Neue Methode zur Glycosid- und Oligosaccharidsynthese - gibt es Alternativen zur Koenigs-Knorr-Methode? - R.R. Schmidt

Hexaethinylbenzol - R. Diercks et al

PHYSIK IN UNSERER ZEIT 17(1986)1
Das Photon - K. Heinloth

Nuklearer Winter und historische Klimakatastrophen

Mikroelektronik - W. Hilberg

PHYSIK IN UNSERER ZEIT 17(1986)2
Lineare und ebene Kristalle - H. Genz

Abbremsung und Einfang von Na-Atomen - J. Fricke

Sychrotron-Strahlung als Moessbauerquelle - G. Fritsch

LA RECHERCHE (1985)171 (in French)
Carnivorous plants - U. Luettge

Biodegradation of hydrocarbons - A. Sirvins; B. Tramier

Niels Bohr and the world's strangeness - B.d'Espagnat

If only neutrinos had mass! - J.J.Simpson

LA RECHERCHE (1985)172 (in French)
NMR imaging - E. Bretan

Controling cholesterol - P.T. Kavanen

Geology of the planet Mars - Ph. Masson

A possible answer to the puzzle of solar neutrinos - M. Spiro

Die in der Technik breit angewendeten radiometrischen Verfahren wie β-Rückstreumessung u. ä. [1, 2], ergeben im vorliegenden Fall keine befriedigenden Ergebnisse. Ein von *Vetter* [3] vorgeschlagenes Verfahren unter Anwendung von ^{14}C erwies sich als technisch zu aufwendig, während eine autoradiografische Methode [4] im vorliegenden Fall zu zeitintensiv ist. Als nutzbare Lösungsvariante ergab sich somit die Anwendung offener, γ-strahlender Radionuklide in Form einer nichtisotopen physikalischen Markierung. Dafür sind Radionuklidgeneratoren auf Grund der guten Verfügbarkeit kurzlebiger Nuklide besonders geeignet [5]. Die in der Regel für medizinische Untersuchungen vorgesehenen Isotope sind jedoch nur bedingt für technische Anwendungsfälle einsetzbar. In Tab. 1 sind Nuklidgeneratorsysteme nach steigender Halbwertszeit des Tochternuklids geordnet, die nach [6] eine Anwendung vornehmlich in der Technik erwarten lassen. Aus Gründen der guten Verfügbarkeit und der wahrscheinlichen Einsetzbarkeit für Aufgaben der Prozeßkontrolle wurden die Untersuchungen mit dem bereits vielfach bewährten $^{99}Mo/^{99m}Tc$-Generator begonnen. Es ist beabsichtigt, die hier dargelegten Versuche mit weiteren in Tab. 1 aufgeführten Systemen fortzusetzen.

Bei den im folgenden beschriebenen Untersuchungen wird dem zu messenden Schichtmedium ein sich homogen verteilendes, strahlendes Nuklid zugesetzt, dessen Reichweite relativ klein ist.

Die mit diesem markierten Substrat erzeugte Schicht sendet damit eine radioaktive Strahlung aus, deren Stärke von einem gewöhnlichen Strahlungsdetektor in geeigneter Abschirmung und in Verbindung mit einem Strahlungsmeß- und Registriergerät bestimmt wird. In Abhängigkeit von der Aktivität des Stoffes sowie den anderen im weiteren zu untersuchenden Einflußfaktoren ist das Meßergebnis ein Maß für das strahlende Schichtvolumen bzw. die Schichtdicke. Es kann also eine direkte Proportionalität zwischen der Schichtdicke und der Anzahl der radioaktiven Zerfälle (Aktivität) und damit der Zählrate erwartet werden.

2. Theoretische Betrachtungen zur Meßgeometrie

Das Radionuklid ^{99m}Tc emittiert eine niederenergetische Gammastrahlung ($E_\gamma \sim 0{,}14$ MeV). Die Meßstrecke enthält das Detektionssystem, bestehend aus einem NaI(Tl)-Szintillator, einer PSEV-Sonde und einem Kollimator-, Blenden- und Abschirmungssystem, sowie die Auswerteeinheit, bestehend aus dem Strahlungsmeßgerät und dem Ergebnisdrucker.

Nachfolgendes Schema soll den Gegenstand der Messungen illustrieren:

Walzengeschwindigkeit $\xrightarrow{f_1}$ Schichtdicke $\xrightarrow{f_2}$ Zählrate

$f_2; f_1$

Messung

Es gilt, aus den gemessenen Zählraten auf die Schichtdicke zu schließen, wobei meßtechnisch nur die Zählraten und die Walzengeschwindigkeiten zugänglich sind. Das heißt, um den funktionalen Zusammenhang f_2 aus dem gemessenen f_2; f_1 zu ermitteln, müssen auf der Grundlage anderer Experimente Annahmen über f_1 vorgenommen werden. Die Funktion f_2 wird von zahlreichen physikalischen Größen beeinflußt. Das sind zum einen die vom Detektionssystem bestimmten Größen, wie die effektive Detektoroberfläche, die Kollimatorwirkung und die relative Lage des Detektorsystems zur Strahlenquelle [7]. Zum anderen sind es die von der Auswerteeinheit beeinflußten, wie der Pegel, die das Signal-Rausch-Verhältnis kontrollierende Verstärkung sowie die Auswahl eines geeigneten Plateaupunktes als Arbeitspunkt. Zusätzlich wirkt die eingesetzte Aktivitätsmenge als experimenteller Parameter.

Das Untersuchungsobjekt sind dünne Schichten, die auf einer rotierenden Walze haften. Da sich der Walzendurchmesser um Größenordnungen von der Schichtdicke unterscheidet, ist eine auf dieses Meßproblem angepaßte optimale Meßgeometrie erforderlich. Nachfolgende Einflußgrößen sollen qualitativ und quantitativ diskutiert werden.

Unter Beachtung der Streueffekte an der Walzenoberfläche sind die Einstellparameter am Strahlungsmeßgerät zu wählen. Eine Zunahme des Meßeffekts ist mit steigendem Verhältnis des erfaßten Schichtvolumens zur Untergrundstrahlung zu erwarten. Diese Untergrundstrahlung wird entscheidend von der geometrischen Form des Flüssigkeitsbehälters und der darin befindlichen Aktivitätsmenge bestimmt. Kennzeichnend für das Energiespektrum dieser Strahlung ist, daß der Anteil der Energiequanten in Umgebung der maximal möglichen Energie ($E_\gamma \sim 0{,}14$ MeV) größer ist als in jenem Energiespektrum, welches sich nach einer Wechselwirkung der Gammaquanten mit der Walzenoberfläche ergibt. In Abhängigkeit von der eingesetzten Aktivitätsmenge ist die Sonde zweckmäßig derart abzuschirmen, daß Untergrund- und Streustrahlung hinreichend stark reduziert werden.

Abb. 1 dient zur Illustration der Abhängigkeit des Meßeffektes von der relativen Lage des Detektorsystems zur Flüssigkeitsschicht. Der abgebildete Schnitt senkrecht zur Achse der rotierenden Walze soll die Tendenzen bei der Optimierung richtig widerspiegeln. Dabei ist a der Abstand des Detektionspunktes P von der Walzenachse, $2w$ der Walzendurchmesser, d die gesuchte Dicke der haftenden Flüssigkeitsschicht, α_0 der die Lage charakterisierende Winkel und α der vom passiven Abbildungselement abhängige Winkelbereich, aus welchem die emittierten Gammaquanten erfaßt werden können. Zur Vereinfachung des Modells wird nur Direktstrahlung betrachtet.

Für die Zählratendichte pro Längeneinheit der Walze $\dot{N}_p$ [$m^{-3}\,s^{-1}$] erhält man

$$\dot{N}_p = C \int_{\alpha_0}^{\alpha_0+\alpha} \int_{w}^{w+d} \frac{q(r', \varphi)\, r'\, dr'\, d\varphi}{r'^2 + a^2 - 2r'a \cos \varphi} \tag{1}$$

wobei C eine dimensionslose Detektionskonstante und $q(r', \varphi)$ [$m^{-3}\,s^{-1}$] die spezifische Aktivitätsmenge der Flüssigkeitsschicht bezeichnet. Unter der Annahme, daß

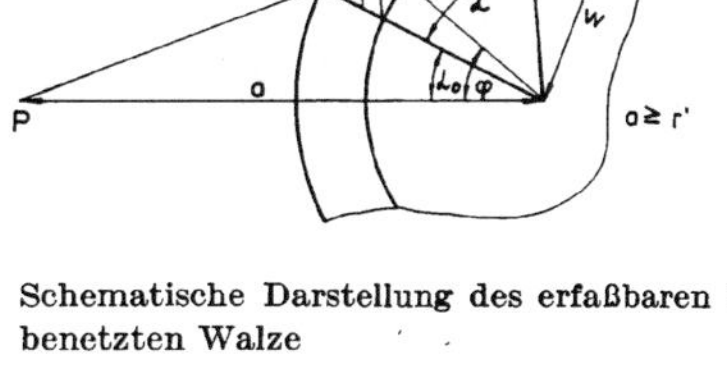

Abb. 1. Schematische Darstellung des erfaßbaren Schichtvolumens der benetzten Walze
(nicht maßstäblich)

die Aktivitätsmenge homogen verteilt ist, liefert (1)

$$\dot{N}_p = Cq \int_{\alpha_0}^{\alpha_0+\alpha} d\varphi \left\{ \frac{1}{2} \ln \frac{(d+w)^2 + a^2 - 2(d+w)\, a \cos\varphi}{w^2 + a^2 - 2wa\cos\varphi} + \frac{\cos\varphi}{|\sin\varphi|} \left[\arctan \frac{(d+w) - a\cos\varphi}{a|\sin\varphi|} - \arctan \frac{w - a\cos\varphi}{a|\sin\varphi|} \right] \right\} \tag{2}$$

Beachtet man die experimentellen Gegebenheiten, wonach die Größen a und w gegenüber d um drei Zehnerpotenzen größer sind, ist (2) nach entsprechenden Reihenentwicklungen und Winkelintegration umschreibbar in

$$\dot{N}_p \sim q\{F(\alpha_0 + \alpha) - F(\alpha_0)\} \tag{3}$$

mit

$$(Fx) = C \left\{ x \frac{d^2}{2w(2w+d)} + \frac{d(dw + w^2 - a^2)}{w|w^2 - a^2|} \arctan \frac{(w+a)^2}{|w^2 - a^2|} \tan\frac{x}{2} + \frac{2d(w^2 + a^2 + dw)}{(2w+d)\sqrt{(w^2+a^2+dw)^2 - a^2\,(2w+d)^2}} \times \arctan \frac{w^2 + a^2 + dw + a(2w+d)}{\sqrt{(w^2+a^2+dw)^2 - a^2(2w+d)^2}} \tan\frac{x}{2} \right\} \tag{4}$$

Infolge weiterer Näherungen, der Vernachlässigung von d gegenüber w und a sowie der Umbenennung $d = 10^{-3}\, d'$, ergeben sich Gleichung (3) und (4) zu

$$\dot{N}_p \sim Cq \left\{ \alpha \cdot \frac{10^{-6}}{4w^2} d'^2 + d' \cdot 10^{-3} \frac{2w}{|w^2 - a^2|} \times \left[\arctan \frac{(w+a)^2}{|w^2 - a^2|} \tan\frac{\alpha_0 + \alpha}{2} - \arctan \frac{(w+a)^2}{|w^2 - a^2|} \tan\frac{\alpha_0}{2} \right] \right\} \tag{5}$$

Gleichung (5) ist zu entnehmen, daß die Zählratendichte $\dot{N}_p$ zum einen quadratisch und zum anderen linear von der Schichtdicke d' abhängt. Das Maß der quadratischen Abhängigkeit ist über den Winkel α steuerbar. Demgegenüber ist der Anstieg im linearen Term sowohl durch die Winkel α_0 und α als auch durch die Entfernung des Detektionspunktes von der Schicht bestimmt. Dieser Einfluß soll Gegenstand weiterer Betrachtungen sein. Der von den Lageparametern abhängige Teil des Anstiegs hat die Struktur

$$f(\alpha_0, \alpha, x) = x \left[\arctan\left(kx \tan \frac{\alpha_0 + \alpha}{2} \right) - \arctan\left(kx \tan\frac{\alpha_0}{2} \right) \right], \tag{6}$$

worin $k = (w + a)^2$ und $x = |w^2 - a^2|^{-1}$ gilt.

Da das Modell gemäß (1) Absorptions- und Streueffekte infolge der Wechselwirkung der Strahlung mit dem Walzenmaterial nicht beinhaltet, kann zur Diskussion nur der Bereich $0 \leqq (\alpha_0 + \alpha) \leqq \pi/2$ herangezogen werden. Die Winkel α_0 und α sind abhängig von der Orientierung des Kollimators sowie dem Durchmesser seiner Öffnung ($2R$).

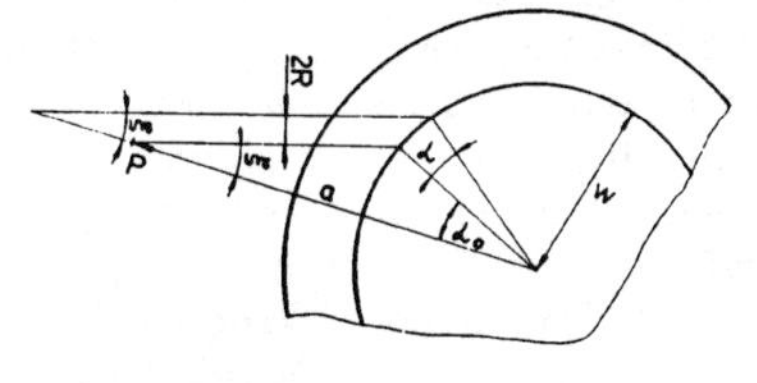

Abb. 2. Prinzipskizze zur Darstellung der Meßgeometrie (nicht maßstäblich)

Entsprechend Abb. 2 wird diese Orientierung durch den Winkel ξ quantifiziert. Es gelten die Zusammenhänge:

$$\sin\xi = w/a \sin(\alpha_0 + \xi) \tag{7.1.}$$

$$\sin(\xi + \alpha_0 + \alpha) = 2R/w + a/w \sin\xi \tag{7.2.}$$

Das Problem der richtigen Positionierung besteht darin, einen optimalen Winkel ξ zu finden, bei welchem die Winkel α und α_0 einen maximalen Wert für das Produkt $d'wf$ ergeben.

Im Bereich $0 \leqq (\alpha_0 + \alpha) \leqq \pi/2$ erhält man bei der Lösung des Extremwertproblems $\alpha(\alpha_0, \xi) \to$ Maximum unter Beachtung der Nebenbedingung (7.1.) mit $\sin(\alpha_0 + \alpha) = -R/w$ und $\sin\xi = -R/a$ keine sinnvolle Lösung.

Nachfolgende Abb. 3 verdeutlicht die Abhängigkeit der Zählratendichte vom Stellwinkel bei vorgegebener Meßanordnung, wobei die dimensionslosen Größen a/w, R/w und kx als Kurvenparameter dienen. Gemäß (5) und (6) charakterisiert das dimensionslose Produkt $d'wf$ den Zählrateneffekt. Die gewählten Kurvenparameter entsprechen praktikablen Meßanordnungen. Es ist darauf hinzuweisen, daß geometrisch ähnliche Meßanordnungen (gleiche Kurvenparameter) zu gleichen Meßresultaten führen. Für die Schichtdicke wurde $d = 1\ \mu m$ ($d' = 1$ mm) gewählt. Der maximale Stellwinkel ξ_{max} ist durch (7.1.) und (7.2.) gegeben.

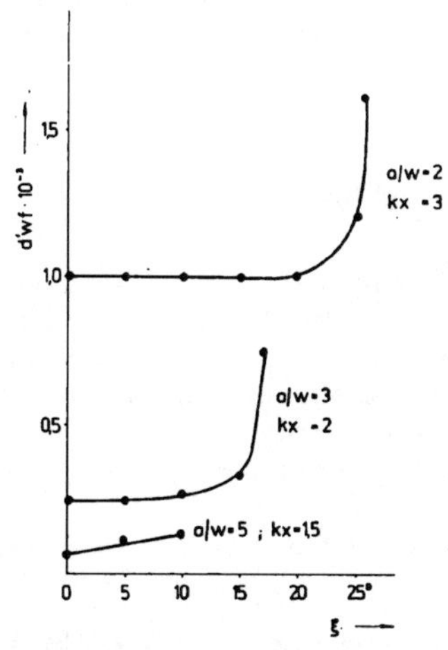

Abb. 3. Orientierungsabhängigkeit des linearen Detektionseffektes

Aus (5), (6) und (7) lassen sich folgende Schlußfolgerungen ableiten:

1. Der dominierende Zusammenhang zwischen der Zählratendichte und der Schichtdicke ist ein linearer, setzt man übliche Verhältnisse für w, a, α_0 und α voraus.
2. Bei hinreichend empfindlichen Detektionssystemen sowie entsprechender Kollimation bzw. Positionierung ist auch die quadratische Abhängigkeit der Zählratendichte von der Schichtdicke meßtechnisch nutzbar.

3. Es liegt bei gegebener Kollimatorgeometrie und gewählten Abstandsverhältnissen kein lokales Maximum für den Winkel α vor, wodurch die Winkel ξ und α_0 bestimmt wären.
4. Der Abb. 3 ist zu entnehmen:

   a) Für die gegebene Meßgeometrie ist der lineare Zusammenhang gegenüber dem quadratischen ($\alpha(d'/w)^2 \sim 10^{-4}$) der dominierende.
   b) Geringfügige Abweichungen von der senkrechten Positionierung des Kollimators ($\xi \sim 0°$) verbessern nicht die Detektion. In Umgebung von ξ_{max} steigt der Detektionseffekt stark an. Dies entspricht einer annähernd tangentialen Anordnung.
   c) Dieses steile Anwachsen verstärkt sich mit abnehmendem Abstand des Detektionspunktes von der zu untersuchenden Schicht (zunehmendes Verhältnis a/w, abnehmendes Produkt kx).

3. Experimente und Ergebnisse

3.1. *Experimentelle Voruntersuchungen*

Um zunächst die prinzipielle Eignung des vorgeschlagenen Verfahrens für die Schichtdickenbestimmung zu untersuchen, und die unter Punkt 1. aufgestellte Hypothese einer direkten Proportionalität zwischen Schichtdicke und Zählrate zu überprüfen, wurden Versuche an ebenen Flächen unter statischen Bedingungen durchgeführt. Auf einer Fläche definierter Größe wurden dabei bestimmte Volumina aufgetragen, die unter der Voraussetzung einer gleichmäßigen Verteilung einer berechenbaren Schichtdicke entsprachen. Da eine Mindestdicke von einem Millimeter erforderlich war, um eine vollflächige Benetzung zu gewährleisten, wurden diese Untersuchungen zunächst im Bereich von 1 bis 12 mm durchgeführt. Gleichzeitig war damit die Vorstellung verbunden, daß Analogieschlüsse für den Mikrometerbereich möglich sind.

Der Versuchsaufbau ist aus Abb. 4 zu erkennen. Unter Berücksichtigung optimaler Geräteeinstellwerte (Pegel, Verstärkung, Kanalbreite) sind damit gute Ergebnisse zu verzeichnen. Die in Abb. 5 dargestellte Funktion $\dot{N} = f(s_D)$ bestätigt die direkte Proportionalität der beiden Variablen. Darüber hinaus erhält man bei Berücksichtigung des Abklingens während der Messungen und der Selbstabsorption einen annähernd linearen Zusammenhang. Der Vergleich mit einer zweiten Meßreihe mit anderer Aktivität ergab, daß sich die Zählrate um den gleichen Faktor ändert wie die Aktivität selbst.

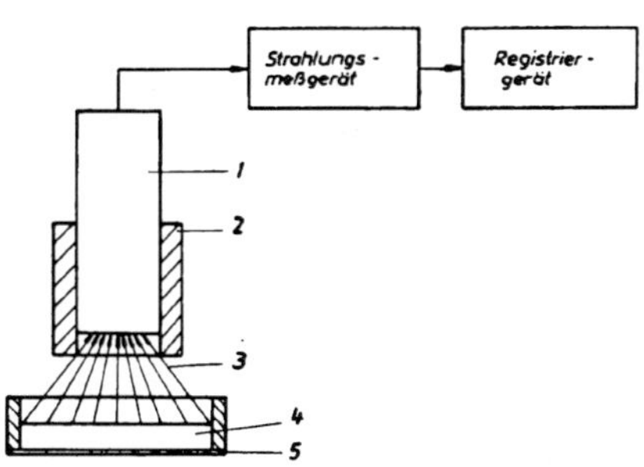

Abb. 4. Meßaufbau für die Vorversuche auf ebenen Proben
1 – Detektor; 2 – Abschirmung; 3 – einfallende Strahlung; 4 – Meßobjekt; 5 – Trägermaterial

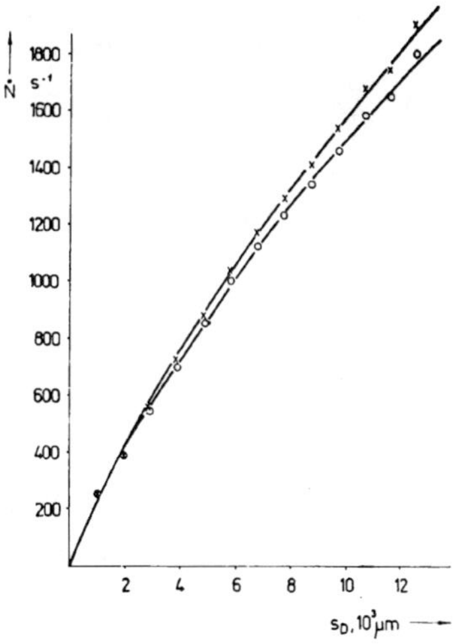

Abb. 5. Zählraten in Abhängigkeit von der Schichtdicke bei ebener Gummioberfläche und geringer Aktivität
○ Werte ohne Zählratenkorrektur; × Werte mit Zählratenkorrektur

3.2. *Untersuchungen an einer rotierenden Walze*

Für die Versuche an einer rotierenden Walze war die Optimierung aller veränderlichen Parameter zwingend erforderlich, da bereits erste Tests ergaben, daß der notwendige Vorratsbehälter der Benetzungsflüssigkeit eine äußerst große Streustrahlung verursachte. Unter Optimierung ist deshalb das Ziel zu verstehen, den von der Schicht ausgehenden Meßeffekt möglichst groß gegenüber den Störimpulsen werden zu lassen bzw. einen meßbaren Unterschied der Zählraten auch für kleine Schichtdickenänderungen zu erzielen.
Da erwartet werden kann, daß sich die Schichtdicke proportional zur Geschwindigkeit ändert, muß damit auch der Zählratenunterschied bei maximaler und minimaler Geschwindigkeit groß sein.

Zur Einschätzung wurde deshalb die relative Zählratenänderung $\Delta\dot{N}_{rel}$ entsprechend Gleichung (8) eingeführt, für die ein Maximum angestrebt wird.

$$\Delta\dot{N}_{rel} = \frac{\dot{N}(v_{max}) - \dot{N}(v_{min})}{\dot{N}(v_{max})} \cdot 100\% \qquad (8)$$

Hierbei sind die Zählraten bei maximaler und minimaler Geschwindigkeit als $\dot{N}(v_{max})$ und $\dot{N}(v_{min})$ bezeichnet. Auf Grund der zu erwartenden sehr kleinen Zählratenunterschiede und der relativ kurzen Halbwertsdauer des eingesetzten Technetium 99m ist eine Zählratenkorrektur entsprechend dem radioaktiven Abklingen vorzunehmen, um die Ergebnisse vergleichbar zu machen.

Als Resultat dieser Optimierung ergaben sich unter anderem der Einsatz eines Bleikollimators von etwa 60 mm Wanddicke und die tangentiale Anordnung des Detektors entsprechend Abb. 6, deren Vorteile bereits theoretisch unter Punkt 2. nachgewiesen wurden. Sie sind außerdem aus einem Vergleich der Abb. 7 und 8 erkennbar. Hier wurden die Energiespektren für v_{max} und v_{min} bei beiden Anordnungen dargestellt. Deutlich sind die wesentlich größeren Zählratenunterschiede für die nach oben gerichtete, tangentiale Einstellung zu erkennen.

Unter sonst gleichen Bedingungen bewirkte diese Anordnung eine Senkung der Impulsrate um etwa 30%, während sich $\Delta\dot{N}$ nur um 3,5% verringerte, was zwangsläufig zu einer Erhöhung der relativen Zählratenänderung führte.

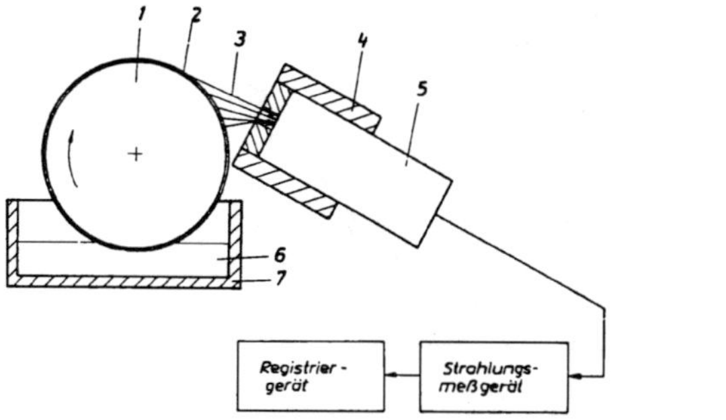

Abb. 6. Tangentiale Meßanordnung an der Versuchswalze

1 – Versuchswalze; 2 – Meßobjekt; 3 – Strahlung; 4 – Abschirmung; 5 – Detektor; 6 – Flüssigkeitsvorrat; 7 – Vorratsbehälter

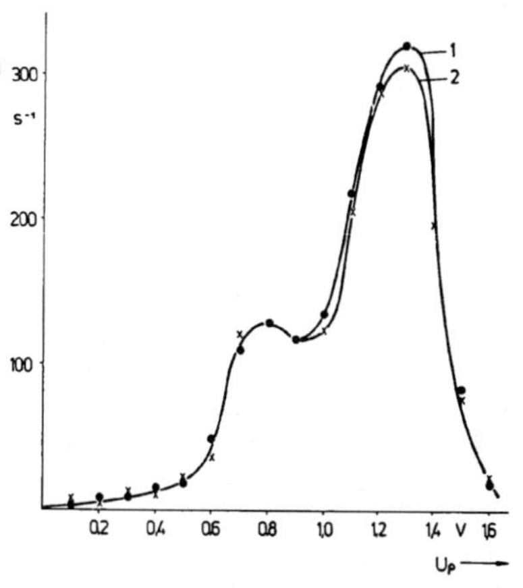

Abb. 7. Energiespektrum bei horizontaler Detektoranordnung

● 1 bei $v = 7{,}0\ \mathrm{m \cdot min^{-1}}$; × 2 bei $v = 3{,}0\ \mathrm{m \cdot min^{-1}}$

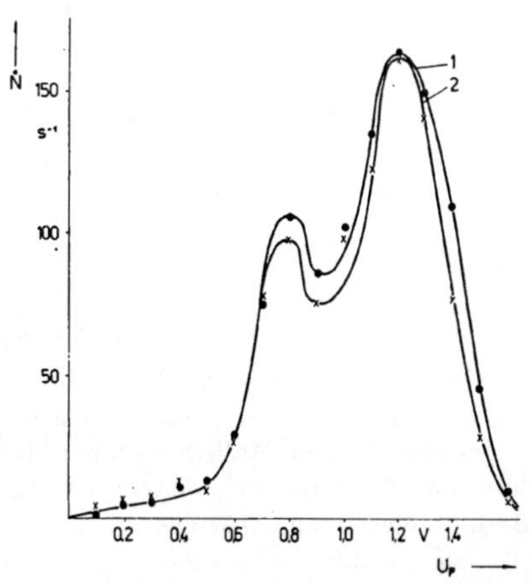

Abb. 8. Energiespektrum bei tangentialer Detektoranordnung

● 1 bei $v = 7{,}0\ \mathrm{m \cdot min^{-1}}$; × 2 bei $v = 3{,}0\ \mathrm{m \cdot min^{-1}}$

Weiterhin erwies sich die vollständige Abschirmung des Detektors durch den Einsatz eines zweiten Kollimators mit 25 mm Wanddicke auf der dem Meßobjekt abgewandten Seite der Sonde als vorteilhaft, da dadurch die den Detektor erreichende Streustrahlung deutlich verringert werden konnte. Durch ihn konnte die Impulsrate auf 61% gesenkt werden, während auf $\Delta \dot{N}$ kein Einfluß festzustellen war. Das bedeutet wiederum eine Erhöhung der relativen Zählratenänderung.

Unter den optimierten Bedingungen wurden abschließend drei Meßreihen zur Überprüfung der Reproduzierbarkeit und Empfindlichkeit des Verfahrens durchgeführt.

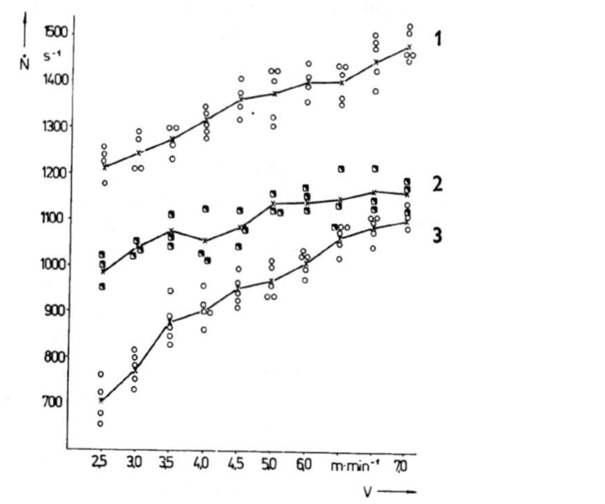

Abb. 9. Zusammenhang zwischen Zählrate und Geschwindigkeit bei verschiedenen Aktivitäten

Versuchsbedingungen: $U = 840$ V; $U_p = 1{,}4$ V diskriminierend; Verstärkung 34 dB; $t = 60$ s

○ ◨ Einzelmessung; × Mittelwert

Dazu erfolgte eine Geschwindigkeitserhöhung um jeweils $0{,}5\ \mathrm{m \cdot min^{-1}}$, die einen Schichtdickenunterschied in der Größenordnung der geforderten Empfindlichkeit bewirkt.

Die Ergebnisse der Versuche mit einer Meßzeit von 60 Sekunden sind in Abb. 9 dargestellt, wobei die Meßreihen 2 und 3 auf Grund der Zählratenkorrektur direkt miteinander vergleichbar sind. Grundsätzlich ist bei allen Meßreihen und den kleinen Schichtdickenänderungen die erwartete Tendenz zu erkennen. Die Notwendigkeit einer relativ langzeitigen Messung von mindestens 5 min (bei der eingesetzten Aktivität) ergibt sich aus der lediglich bei Versuchsreihe 2 mit 3 min Meßzeit auftretenden entgegengesetzten Tendenz.

Als Ursache für die deutlichen Abweichungen zwischen den Kurven 2 und 3 (bis 30%) kann folgendes angenommen werden:

1. Die Reproduzierbarkeit der Schichtdicke selbst kann in gewissen Grenzen schwanken.
2. Geringfügige und unvermeidliche Verschiebungen bei der Einstellung des Arbeitspunktes um etwa 0,5% führen bereits zu Zählratenabweichungen von 15 bis 25%.

Diese Abweichungen erlauben es nicht, für den hier untersuchten Meßaufbau und die gegebenen Bedingungen von einer reproduzierbaren Messung zu sprechen, so daß sich eine weitere Stabilisierung des Verfahrens erforderlich macht.

4. Zusammenfassung

Generell ist eine Schichtdickenbestimmung mit dem vorgeschlagenen Meßprinzip möglich, wie das durch die statischen Vorversuche auf ebenem Untergrund nachgewiesen wurde. Die größten Probleme entstehen durch Streustrahlungen des Meßgutes im Vorratsbehälter oder an anderen Stellen mit Anhäufungen von Aktivität. Trotzdem war unter optimierten Bedingungen eine Zählratenänderung von 15 bis 18% an der Tauchwalze erreichbar. Zur Verbesserung der Dynamik der Messung ist es sinnvoll, mit sinkender Schichtdicke die Aktivität proportional zu erhöhen.

Während das Verfahren in der vorliegenden Form füt viele Anwendungsfälle einsetzbar ist, macht sich besonders für die Bestimmung sehr dünner Schichten und die Beeinflussung durch die Meßgeometrie eine weitere Verbesserung erforderlich, die im Einsatz anderer Nuklidgeneratorsysteme mit höheren γ-Energien der emittierten Strahlung gesehen wird [8].

Eingegangen am 10. Dezember 1985

Literatur

[1] *E. Broda, T. Schönfeld,* Die technischen Anwendungen der Radioaktivität, Leipzig: Deutscher Verlag für Grundstoffindustrie 1962

[2] *L. G. Erwall, F. Forsberg, K. Ljunggren,* Radioaktive Isotope in der Technik, Braunschweig: Verlag Vieweg & Sohn 1968

[3] *R. Vetter,* Entwicklung einer ^{14}C-Strahlenquelle hoher Aktivität zur Messung von dynamischen Flächenmasseänderungen nach dem Rückstreuprinzip, Isotopenpraxis 18 (1982) 247

[4] *H. Baumbach, W. Sobek,* Autoradiografische Messung von Schichtdicken im Mikrometerbereich auf polygrafischen Materialien, Beitrag zum 12. internationalen Symposium Autoradiografie, Harkany, VR Ungarn, 1981

[5] *K. H. Lieser,* Kernchemie, 2. Aufl., Weinheim: Verlag Chemie 1980

[6] *E.-J. Langrock,* Dissertation B, Leipzig: Technische Hochschule 1986

[7] *F. Häußler,* Betrachtungen zum optimalen Kollimatoreinsatz für die Analyse von Gammastrahlungs- bzw. Neutronenfeldern Teil I, Wissenschaftliche Zeitschrift der Technischen Hochschule Leipzig, im Druck

[8] *E.-J. Langrock, J. Stopporka, F. Häußler, H. Baumbach,* Verfahren zur Bestimmung der Schichtdicke von Flüssigkeitsfilmen, DDR-Patentanmeldung WP G 01 B 275406

Isotopenpraxis 22 (1986) 10, S. 361 bis 368

Cell Membranes in Radiation Injury*)

G. J. Köteles

("Frederic Joliot-Curie" National Research Institute for Radiobiology and Radiohygiene Budapest, Hungary)[1]

The introductory paper to the pathophysiology session of the 19th Annual Meeting of European Society for Radiation Biology, 1985, outlines cell membrane-related phenomena elicited by lower doses of low LET radiations than the cell killing one. Through micromorphological appearances, functional activities like that of receptors and binding sites, early appearing temporary alterations are demonstrated. It is suggested that the injuries of cells, cell-to-cell and cell-to-extracellular matrix contacts, the disabled conditions of surviving cells contribute to the development of non-stochastic effects in tissues.

Der Einführungsvortrag der Sektion Pathophysiologie auf der 19. Jahrestagung der Europäischen Gesellschaft für Strahlenbiologie 1985 behandelt Zellmembranphänomene, wie sie im Bereich niedrigerer Dosen einer niedrigen linearen Energieübertragung (LET) eintreten als im Bereich einer Zellabtötung. Durch mikromorphologische Erscheinungen und durch funktionelle Aktivitäten, wie sie bei den Rezeptoren und Bindungsorten auftreten können, ist es möglich, frühe, zeitweilige Veränderungen zu erkennen. Es wird vorgeschlagen, daß die Zellschäden (Zelle-zu-Zelle und Zelle-zu extrazellulären Matrix-Verknüpfungen) zerstörende Bedingungen auch auf die überlebenden Zellen ausüben und zum Fortschreiten nicht-stochastischer Effekte in Geweben beitragen.

Keywords

cell membranes; irradiation; LET; morphological changes; radiation injuries; receptors; tissues; ultrastructural changes

Introduction

I am honored to be invited to present an introductory paper at the session of pathophysiology of the 19th Annual Meeting of the European Society for Radiation Biology. A review of recent aspects of radiation pathophysiology – as it has been asked – would be an immense task as radiation injury is a complex phenomenon and the whole building of our contemporary science is still far from revealing all of its details. Due to the broad coverage of this discipline, it is not surprising that in the recent years the reader can hardly find comprehensive reviews on this topic, the last ones were listed by *Zimmer* [1]. Therefore, I want to confine my presentation to some basic elements of radiation injury at cellular, intercellular and tissue level and to observe these problems most of them still to be solved in perspectives. In the followings, few examples of radiation-induced cell membrane-related phenomena, the cell-to-cell and cell-to-extracellular matrix contacts will be presented and through the example of skin injuries some notes will be made concerning cellular contributions to non-stochastic effects on tissues.

Cell membrane-related alterations

The progress in membrane research has lead to the extension of relevant radiobiological studies [2–7]. The investigations aimed mainly to reveal the role of membranes as possible critical radiosensitive targets. From the data available, a contributory role can be attached to the membranes in shaping the fate of cells being it either death or regenerations, as the structural integrity of the membrane seems to be essential for repair performance [8–11],

*) Invited paper presented at the 19th Annual Meeting of the European Society for Radiation Biology, August 1985, Prague, Czechoslovakia

[1]) Address: H-1775 Budapest, P.O. Box 101, Hungary

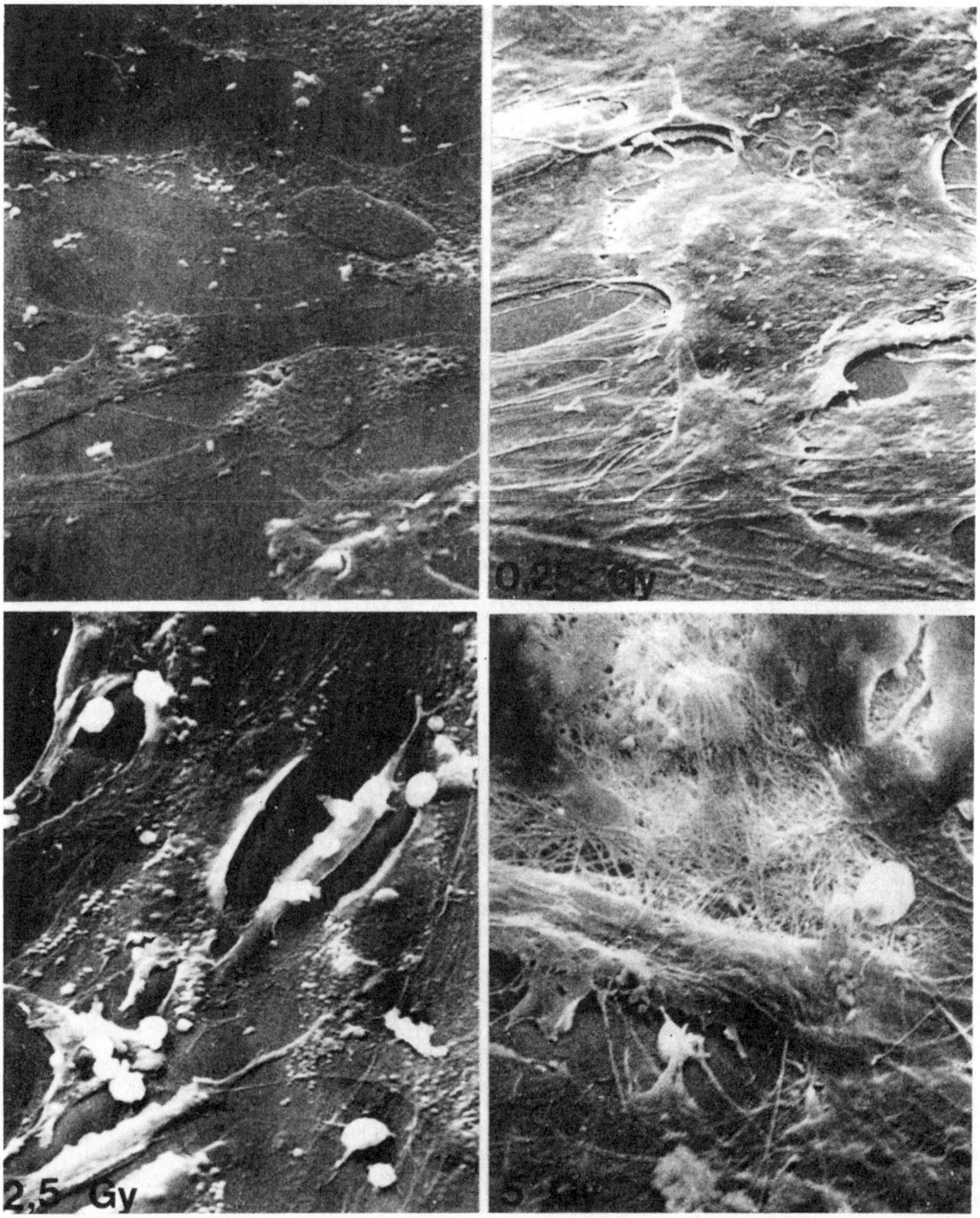

Fig. 1. Scanning electron microscopic appearance of primary human embryonic fibroblasts. Unirradiated control (C: 1200 ×), and cell cultures 30 min after x-irradiation with 0·25 Gy (3800 ×), 2.5 Gy (1200 ×) and 5 Gy (2000 × magnification).

membrane specific drugs like chlorpromazine and procaine inhibit postirradiation repair processes [12].

When dealing with cellular pathophysiology, however, not only cell death has to be considered as end-point, but a series of temporarily existing or long-lasting alterations of cellular structure — like membranes — and functions at molecular or supra-molecular levels. The following examles point to effects after relatively low doses, when the minimum eliciting dose or the D_0 value for the effects are lower than the D_0 value of cell killing.

The *micromorphological appearance* of x-irradiated primary human embryonic fibroblasts shows considerable excitement of plasma membrane (Fig. 1). In the unirradiated culture the cell layer is confluent, the edges of cells adhere tightly to the substrate, elevation of cell edges, i.e. rufflings are seen rarely, the shapes of nuclei are well outlined, at a few sites minor extensions are visible. Shortly after irradiation with 0,25 Gy, the cell edges elevated, many filopodia and lamellipodia developed. The cell-to-cell contacts as well as the cell-to-substrate contacts became loosen. 2·5 Gy induced similar alterations, the detachment of cell edges were more expressed, some cells took up spindle- or rounded forms. After 5 Gy many cells detached fully and only a fibrous extracellular matrix remained at their original places [13].

These micromorphological changes of fibroblasts are similar to those observed on lymphocytes [14, 15]. *Ehrlich* ascites tumor cells [15a], human carcinoma cells [15], glial cells [16] and lymphoma cells [15, 17]. These alterations are early appearing and after relatively low doses they are reversible. The examples mentioned point to an augmented activity of plasma membranes of irradiated cells in ruffling and endocytosis. This might mean increased uptake of nutrients, internalization and replacement of injured membrane segments, increased activity in cell motion. Further question is whether the radiation-induced new formations have the same composition and function as the average membrane segments.

We have studied the *regional differences of cell surface* by labelled ligands. The polycationized ferritin can make the negatively charged surface areas visible by electron microscopy [18]. These charges carried by glycoproteins, glucosaminoglycans play role in building up recognition and response areas, various receptors, conjunctions, binding sites for autologous and heterologous molecules, virus particles, etc. Upon ionizing radiation the amount of negative charges available on the surface drops within a minute after x-irradiation by alrady 0·25 Gy, then recovers within an hour (Fig. 2). The inserts in the figure serve to visualize the basis of semiquantitative evaluation of ferritine grain numbers on cuts perpendicular to the plane of cell culture on 100000× magnification [19]. Beside the number of anionic sites the charge densities of the sites is also decreasing as shown on Fig. 3. This assessment could be done on cuts tangential to the plane of cells. Here again, the preirradiation level turned back in an hour [20]. On V-79 chinese hamster cells similar transient decrease of binding of colloidal iron hydroxide was found [21].

Regional differences of surface areas are known for liver, stomach and epithelial cells. We have demonstrated that even in cultured fibroblasts, the negatively charged areas are located on the apical and lateral surfaces, while the basal one is free from the cationized ligand [18]. The Fig. 4 demonstrates that this distribution is not influenced by irradiation. The decrease of anionic sites at cell-to-cell contacts point to the loosening of these contacts. The molecules responsible for surface charges are varied from cell to cell. While in lymphocytes they are due mostly to neuraminic acid, in fibroblasts the anionic sites are sensitive mostly to hyaluronidase and less to neuraminidase [20]. The radiation-induced new formations, the cytoplasmic extensions are richer in cationized ferritin than the cell body. It is interesting to note that the rest of surface charges after radiation-induced decrease diminishies further upon neuraminidase in a much less extent than with hyaluronidase suggesting a close contact of cell surface with extracellular matrix (Fig. 5).

It can be concluded that regional differences are existing on cell membranes, certain features are influenced by irradiation like the density and amount of charges as well

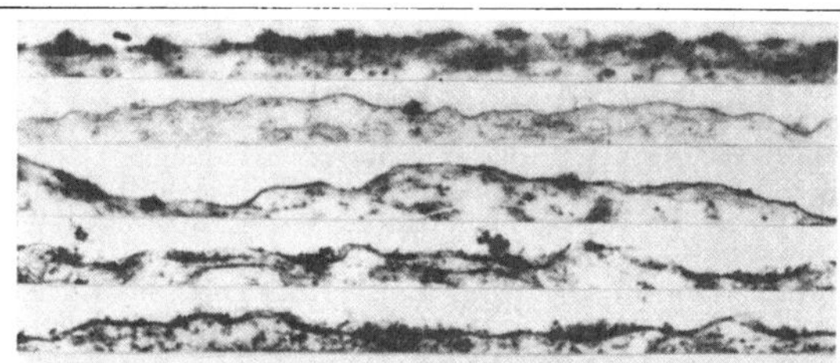

Time after irradiation	Percent of the total free surface
unirradiated control	42.00 ± 14.89
1 min	25.01 ± 11.08
10 min	24.43 ± 9.37
1 h	58.80 ± 10.87
4 h	56.47 ± 14.83

Fig. 2. Cationized ferritin (CF) binding of primary human embryonic fibroblasts at various times after x-irradiation with 2·5 Gy. The extent of CF coverage on cell surfaces was estimated on perpendicular sections of cell cultures as seen on micrographs. Insert magnification: 34500 ×.

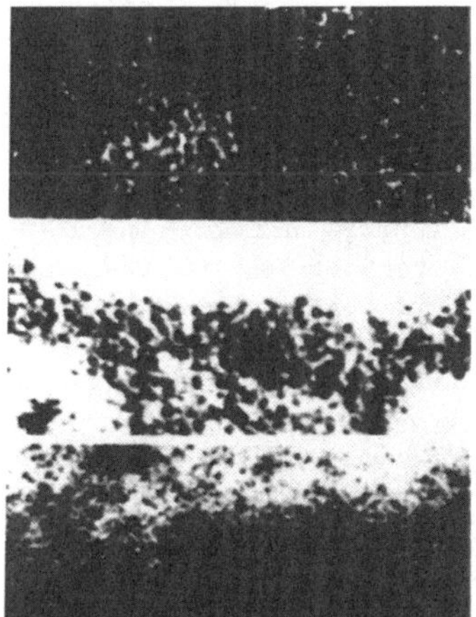

Time after irradiation	Number of particles per µm²	
unirradiated control	4030 ± 250	100
1 min	2250 ± 270	56
–10 min	1660 ± 121	41.1
30 min	2970 ± 143	73.6
1 h	3830 ± 235	95
–4 h	4210 ± 276	104.4
24 h	3940 ± 219	98.2

Fig. 3. Density of cationized ferritin bound within negatively charged binding sites on primary human embryonic fibroblasts at various times after x-irradiation with 2·5 Gy. The number of CF particles was counted in tangential sections through cell cultures as seen on micrographs of 100000 × final magnification at 50 × 1 µm² area per sample. Insert magnification: 100000 ×.

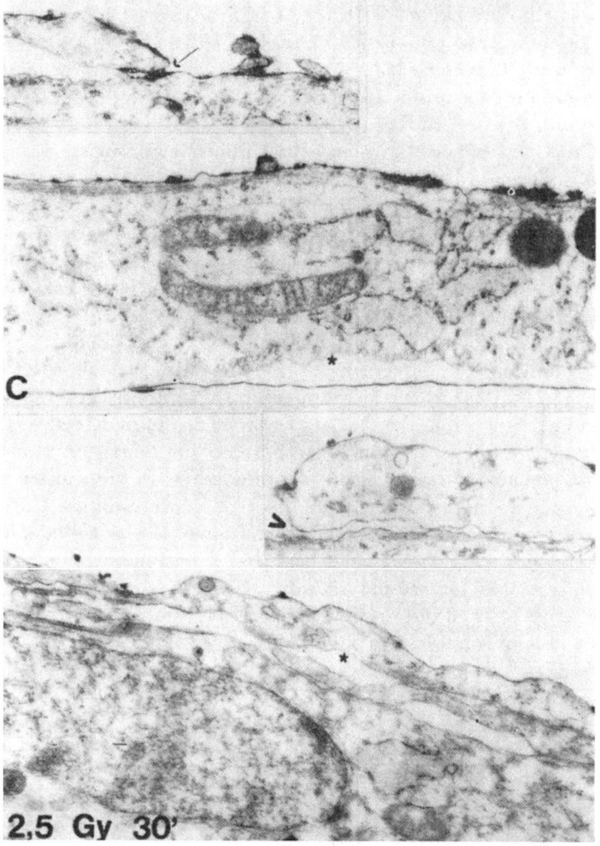

Fig. 4. Localization of cationized ferritin binding sites at various surfaces of primary human embryonic fibroblasts. Unirradiated control (C) and cultures 30 min after x-irradiation with 2·5 Gy. The arrows (→, >) point to cell-to-cell contacts. The asterisks label the basal surfaces of cells. Magnifications: 46000 ×.

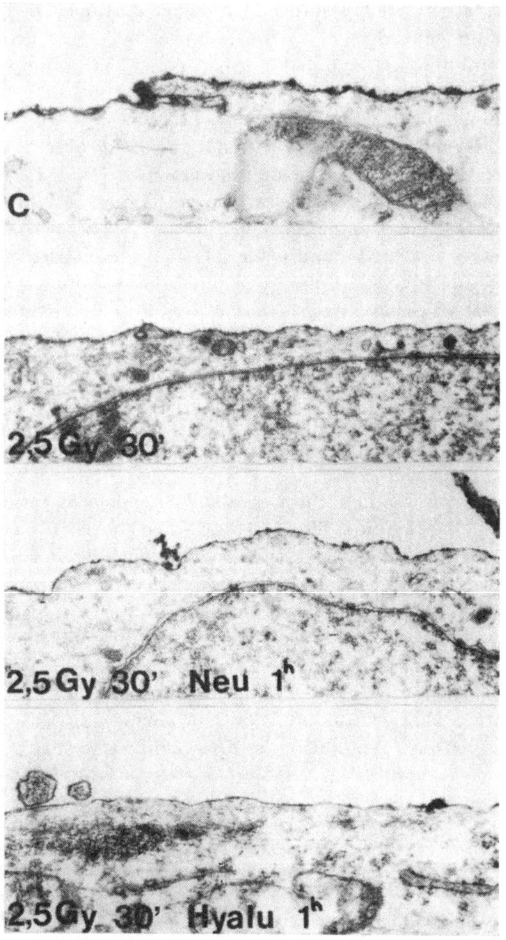

Fig. 5. Cationized ferritin binding at the surfaces of primary human embryonic fibroblasts. Unirradiated control (C) and cultures 30 min after x-irradiation with 2·5 Gy, cultures after digestions with neuraminidase (0·5 U/ml, pH 6, 37 °C) or hyaluronidase (6000 U/ml, pH 6, 37 °C) 30 min following x-irradiation with 2·5 Gy. Magnification: 46000 ×.

as the cell-to-cell and cell-to-matrix contacts, while others are not like the charge distribution among the apical, lateral and basal surfaces.

Behind the micromorphological alterations there are structural derangements and functional changes of *membrane-related activities*. A few examples are listed on Tab. 1 concerning various types of cells, various functional areas like receptors, binding sites for lectins, virus-particles, rosette forming surfaces and the so-called "homing" properties of lymphocytes pointing to cell-to-cell contacts, the lectin-induced blastogenesis of lymphocytes as the first phase of this process is also membrane-bound. It can be seen that the dose values eliciting already alterations and the D_0 values where available and determinable for certain effects are rather low compared to the D_0 values of killing of the same cell types. The latter data are shown on the Tab. 1 only if the authors compared these in the same experiments. Further similar comparative and functional investigations are needed to assess the radiosensitivities of various membranes as compared to other targets.

The *biochemical background* for normal function of plasma membrane is more and more known [37]. Data are almost lacking, however, on the way how energy deposition might disturb individual components or the functional integrity of membranes in terms of microdosimetry when spatiotemporal physical events are projected over biological structure and function [38]. This approach is promising especially in those cases when a certain function is attached to the cooperation of various protein and lipid molecules e.g. in the case of signal transmission process on lymphocytes upon lectin stimulation [39]. We have tested one essential factor for this system, the breakdown and incorporation of arachidonic acid in the membrane. It could be observed that upon x-irradiation of ^{14}C-prelabelled cells only a part of this radiosensitive polyunsaturated fatty acid (PUFA) was degraded (Fig. 6), while pulse labelling after irradiation indicated a slow, delayed uptake until full regeneration (Fig. 6). It follows from the data presented that the regeneration at various levels lasts for different lengths of time: the surface charge distribution turns back quickly to preirradiation level, while the molecular and cellular regeneration, e.g. that of coordinated functions is slower.

Due to the relatively high radiosensitivity of PUFAs, for long time it was thought that the membrane radiosensitivity is primarily the consequence of lipoxigenation of PUFAs. Interesting works were made recently demonstrating that there is no direct correlation between the PUFA content of plasma membrane and the radiosensitivity of cells as far as sublethal repair of DNA and cell survival are concerned [40]. It was also suggested that the UFA peroxidation might not be the only factor for membrane damage [41]. Thus the radiosensitivities of various membrane components seem to be modified by molecular in-

Tab. 1. Radiation effects on certain membrane-related cell activities

Cell type	Effect	Minimum dose tested eliciting the effect, Gy	D_0	Reference
Receptors				
Murine B lymphocyte	Temporary disappearance of IgG receptors	0·10		[22]
Murine mast cells	Depression of degranulation due to decreased Fc availability	0·10		[23]
Murine platelets	Increase of conA binding	0·25		[24]
Murine lymphocytes	Increase of conA binding	1		
Murine erythrocytes	Increase of conA binding	2		
Human platelets	Increase of conA binding	0·1		[24a]
Human lymphocytes	Increase of conA binding	0·5		
Human erythrocytes	Increase of conA binding	2		
Human fibroblasts	Increase of conA binding	0·9		[25]
	Decrease of cationized ferritin binding	0·25		[18, 19]
Monkey kidney	Increase of poliovirus attachment and penetration	2·5		[26]
Rosette formation				
Human B lymphocyte	Decrease of SmIg and Fc, inhibition of SmIg capping	0·5		[27]
	decrease of EAC and ME rosette formation	1		[28]
Human T lymphocyte	Temporary increase, then decrease of E-rosette formation	1	4·4 – 14·3, survival: 30 – 40	[28]
Human T lymphocyte	Decrease of E rosette formation	0·5		[27]
Human T lymphocyte	Decrease of E rosette formation	0·1		[29]
Human T lymphocyte	Decrease of E rosette formation	0·5		[29]
Human T lymphocyte	Decrease of E rosette formation	0·38		[30]
Human T lymphocyte	Decrease of E rosette formation	1	2	[31]
Rat thymus cells	Decrease of autologous rosette formation	0·5	2·4	[32]
			piknosis: 2·4	
Homing				
Murine lymphocytes	Averted from lymph nodes to liver and spleen	1		[33] [34]
Lectin-induced blast formation				
Human lymphocytes	Increased response	0·13	2·6 – 3·0	[35]
	Inhibition	0·5		[35]
	Inhibition	1		[36]

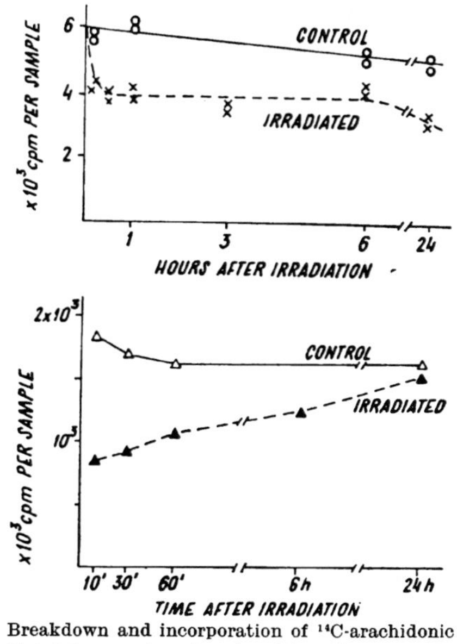

Fig. 6. Breakdown and incorporation of ^{14}C-arachidonic acid in x-irradiated human lymphocytes. Top: Decrease of amount of ^{14}C-arachidonic acid after 2 Gy x-irradiation of prelabelled cells, in vitro. Bottom: Incoporation of ^{14}C-arachidonic acid during 10 min pulses after various times following 2 Gy x-irradiation, in vitro.

teractions in cellular structures and by the intrinsic antioxidant capacities of cells.

The micromorphological importance of integrant proteins was pointed out recently, when shown that the radiation-induced changes of membrane proteins in erythrocytes induce the transformation of cellular shape from discocyte to echinocyte [42].

Among the factors leading to membrane perturbation, the damage of *cytoskeleton system* is probably involved. It was, however, described that in the dose-range provoking the structural and functional derangement, the contribution of assembly of microtubular proteins might have only insignificant effect [43]. In comparison, the microtubules proved to be more radiosensitive than microfilaments in macrophages [44].

Cell-to-cell and cell-to-matrix contacts

The cell-to-cell and cell-to-matrix contacts have decisive role in organization of tissues. Recent studies revealed multiple mechanisms and molecular involvement in cell surface interactions with extracellular matrix [45, 46]. In the interaction several types of molecules participate like glycoproteins, phospholipids, glycolipids and extracellular material like fibronectin, laminin, collagen, elastin, heparan sulphate proteoglycans. The extracellular matrix influences cellular activities like movement [47].

The radiopathological importance of such studies would be multiple, e.g. in developmental radiobiology, pathology of wound healing, inflammation, tumor invasion and nonstochastic effects of ionizing radiations. It was also suggested that cell-to-cell contacts are modifying factors in radiosensitivity and they influence repair of radiation damage [48]. Among the few available radiobiological data of relevance, the aforementioned phenomena gave already micromorphological and functional evidences on the loosening of cell-to-cell contacts upon ionizing radiation. Concerning the cell-to-matrix connections it was demonstrated that in the fibrosis of lung the loss of anionic sites at the alveolar and capillary basal laminar membrane occured within an hour and lasted for 4—12 weeks following 5 Gy producing no other histologically detectable signs of injury as well as after doses producing fibrosis and pneumonitis, 9 and 13 Gy, respectively [49]. These findings are in good agreement with our observations as mentioned concerning the hyaluronidase sensitive anionic sites on the surface of fibroblasts which support the view that cell surface is determined by cellular and extracellular compounds, on the other hand radiation loosens the cell-to-matrix contacts.

Tissue injuries: the skin

In the pathophysiology of radiation injuries great concern has developed which is reflected also in the ICRP Publication 41 "Nonstochastic Effects of Ionizing Radiations" [50, 51]. The nonstochastic effects are defined as types of damage due to collective injury to a certain number or proportions of cells, there will be a threshold below which the number or proportions of cells effected is insufficient for nonstochastic injury to occur. This threshold depends, per se, on the choice of level of injury or the sensitivity of detection. It is stated that "In the pathogenesis of most — if not all — nonstochastic effects, the killing of cells play a crucial role. Although other mechanisms may be involved". At our present session I would like to put an emphasis on the last sentence of this quotation. Cellular pathophysiology — as a few examples were given — suggests that cells are effected by lower doses than the killing one, they survive and function, but this might mean unhealthy, disabled conditions of cells. These might have consequences on the performance of tissues. The tissue injury depends on the cellular compositions and the individual radiosensitivities of component cells, therefore, we have to know these including the interactions of various cell types.

Out of the several known nonstochastic effects the injury of connective tissue and blood vessels merit a special attention as being common in most organs. The skin includes also both and it has a significant role either in radiation accidents or in radiotherapy. The multicoloured symptomatology of skin injury includes three phases of reactions, the durations of which extend from hours to years (Tab. 2). For the symptoms of the first wave, — the transient erythema, moist and dry desquamations, temporary epilation — the injuries of epithelial germinal cells, the capillary endothel cells and the hair follicles are responsible. For the second wave, the endothelial cells of dermal blood vessels and the fibroblasts are the most sensitive critical targets. The nonstochastic late consequences are the dermal atropy, fibrosis, permanent epilation, pigmentation. The approximate threshold doses are known for a few symptoms from clinical experience. A more precise approach would be, however, the determination of cell survival kinetics of various cell types. The Table 3 demonstrates a few example: the epidermal basal cells are the most sensitive, then they are followed by the endothelial cells and fibroblasts. It is also seen on the Tab. 3 that there are subpopulations of differing radiosensitivities in epidermal cells, and also individual and species differences of radiosensitivities of fibroblasts. The data give only an impression on the dose-ranges under in vitro conditions or in isolated in vivo systems. For the assessment of "threshold" doses on cellular reactions of various cells comparative

Tab. 2. Radiation-induced reactions of the skin. Symptoms and consequences

Reaction phases	First wave	Second wave	Late effects
Time of appearance	Hours	Days, weeks	Month, years
	Erythema 6 – 8 Gy* 30 Gy** Desquamation moist, dry Temporary epilation 3 – 4 Gy Necrosis	Erythema 20 – 30 Gy Ulceration Necrosis Reduction in normal wound healing	Dermal atrophy – Reduction of elasticity – Fibrosis Teleangiectasia Permanent epilation 7 Gy* 50 – 60 Gy** Pigmentation
Critical target cells	Germinal cells of epithelial layer Capillary endothel, Hair follicles	Endothel cells of dermal blood vessels, Fibroblasts Sebaceous glands, Hair follicles	

* Single irradiation
** Fractionated irradiation

Tab. 3. Comparison of cell survival kinetics data of various cells in skin following low LET irradiations

Cell type	D_0	D_q	n	Reference
Epidermal basal				
Human	0·74		10	[52]
	0·88		16	
Murine	1·2			
Swine	2·7			[53]
	3·4			
	5·7			[54]
	16·2			
Endothelial				
Capillaries (Human, bovine, murine)	1·6 – 2·0	1·8 – 2·9		[55]
Vascular density	1·7	3·4	7	[56]
Vascular growth in subcutis	1·7			[57]
Fibroblasts				
Human	1·36		0·94	[58]
	1·6		1·52	
	1·7		1·0	[59]
	2·4		2·9	
	2·5		4·7	
Swine	4·0			

studies would be needed and performed simultaneously under the condition of tissue organization. Recently investigations were initiated concerning the dependence of skin injuries on radiation quality as in radiation accidents with β-sources the particles of different energies reach different depths [60, 61]. This is a promising approach to correlate symptoms and threshold doses of critical target cell populations. In the recent years considerable progress was made in the radiation pathophysiology of capillaries and blood vessels which contribute basically to late effects of radiation [53, 54, 62, 63, 64].

Finally, to include a practical problem on radiation injury of tissues, I mention that the thermographic investigation on irradiated tissues is a very sensitive technique. In a case of an injured hand the clinical signs — erythema and phlyctena — of the second phase of injury were seen on much smaller areas than the injury seen by thermography. The thermographic analysis was also positive after the second phase has been terminated. These observations — together with those from our running experiments on mini-pigs according to which the thermal map is changed before the appearance of clinical signs — point also to the early appearing and long-lasting reactions of blood vessels. In addition, such observations support the view that observation on nonstochastic effects depend on the sensitivity of technique for detection.

Conclusions

The intention of this paper was to stress that

- cellular pathophysiology needs more attention in dose ranges below the killing one, especially on the degree of abilities of cells to function,
- cell-to-cell and cell-to-matrix contacts may influence the extent of injury and the fate of cells being it either full or partial regeneration or cell death,
- the understanding of pathophysiology of nonstochastic effects require simultaneous investigations on time and dose dependence of different cell populations within a tissue and on their interrelationships.

Acknowledgements

The interest and support of Professor Dr. *L. B. Sztanyik*, Director-General of the Institute, the scientific collaborations of Drs. *T. Kubasova* and *Z. Somosy*, the technical assistance of Mrs. *E. Hrisztodulakisz* and *E. Fodor* are highly acknowledged.

Received November 27, 1985

References

[1] *K. G. Zimmer*, Advances in Radiation Biology **9** (1981) 411
[2] *D. K. Myers*, Advances in Biology and Medical Physics **13** (1970) 219
[3] *D. H. F. Wallach*, Biomembranes **5** (1974) 213
[4] *G. Patrick*, The effects of radiation on cell membranes, in: "Mammalian Cell Membranes", ed. by *G. A. Jamieson, D. M. Robinson*, Vol. 5, London: Butterworths 1977, 72
[5] *G. J. Köteles*, Atomic Energy Review **17** (1979) 3
[6] *G. J. Köteles*, Radiation and Environmental Biophysics **21** (1982) 1
[7] *J. C. Edwards, D. Chapman, W. A. Cramp, M. B. Yatvin*, Progress in Biophysics and molecular Biology **43** (1984) 71
[8] *J. Szumiel*, Advances in Radiation Biology **9** (1981) 281
[9] *H. Schorn, U. Bertsche, M. Nüsse, P. E. Bryant*, Radiation Environmental Biophysics **19** (1981) 301
[10] *H. Schorn, U. Bertsche, M. Nüsse*, The influence of membrane-active drugs on the radiosensitivity of Ehrlich ascites tumor cells. Proceedings of International Congress of Radiation Research, Amsterdam 1983, B3 – 29
[11] *W. T. Nolan, J. E. Thompson, J. R. Lepock J. Kruuv*, Internat. J. Radiation Biology **39** (1981) 195
[12] *S. Yonei, C. Y. Mitsutani, T. Todo, O. Niwa*, Modification of radiosensitivity of bacterial and mammalian cells by membrane-specific drugs. In: "Modification of Radiosensitivity in Cancer treatment", Academic Press Japan 1984
[13] *Z. Somosy, T. Kubasova, G. J. Köteles*, J. Radiation Research **24** (1983) 109
[14] *T. Yamada, H. Ohyama*, J. Radiation Research **21** (1980) 190
[15] *S. Chandra, S. Stefani*, Internat. J. Radiation Biology **40** (1981) 305
[15a] *S. Skog, V. P. Collins, B. Ivarsson, B. Tribukait*, Acta Radiologica Oncology **22** (1983) 151
[16] *H. Hamberg, U. Brunk, J. L. E. Ericsson, B. Jungs*, Acta Pathologica Microbiologica Scandinavica, Sect. A, **86** (1978) 487
[17] *T. Yau*, Scanning Electron Microscopy **4** (1981) 47
[18] *Z. Somosy, T. Kubasova, G. J. Köteles*, Cell Biology International Reports **8** (1984) 407
[19] *Z. Somosy, T. Kubasova, G. J. Köteles, J. Kovács*, Electronmicroscopic studies on radiation-induced alterations of negative surface charges of human fibroblasts. Electron Microscopy 1984. EUREM 1984, Vol. 3, 1867 – 1968
[20] *Z. Somosy, T. Kubasova, G. G. Ecsedi, G. J. Köteles*, Internat. J. Radiation Biology, **49** (1986) 969
[21] *J. R. Subjeck, S. W. Hui, J. R. Johnson*, British J. Cancer **41** (1980) suppl. IV. 159
[22] *H. Folch, M. I. Guarda, F. Ojeda*, Immunology Letters **5** (1982) 337
[23] *M. I. Astorquiza, F. Ojeda*, IRCS Medical Science **13** (1985) 30
[24] *T. Kubasova, L. P. Varga, G. J. Köteles*, Part I, Internat. J. Radiation Biology **40** (1981) 175
[24a] *T. Kubasova, G. J. Köteles, L. P. Varga*, Part II, Internat. J. Radiation Biology **40** (1981) 187
[25] *G. J. Köteles, T. Kubasova, L. Varga*, Nature **259** (1976) 507
[26] *G. J. Köteles, T. Kubasova, Z. Somosy, L. Horváth*, Derangement of cellular plasma membranes due to non-lethal radiation doses. In: "Biological Effects of Low-Level Radiation", International Atomic Energy Agency, Vienna 1983, 115 – 128
[27] *A. Facchini, N. M. Maraldi, S. Bartoli, A. Farulla, F. A. Manzoli*, Radiation Research **68** (1976) 339
[28] *C. Szczylik, W. Wiktor-Jedrzejczak*, Internat. J. Radiation Biology **39** (1981) 253
[29] *S. A. Birkeland*, Transplantation **29** (1980) 23
[30] *F. Ojeda, M. Inés Guarda, D. Moraga, M. I. Astorquiza, H. Folch*, Effect of ionizing radiation on lymphocyte membranes, "Cell Membrane Probes as Biological Indicators in Radiation Accidents", Internat. Atomic Energy Agency, Vienna 1982, IAEA-TECDOC-273, 57 – 76
[31] *S. Albrecht*, Strahlenempfindlichkeit von Lymphozyten in vitro. Staatliches Amt für Atomsicherheit und Strahlenschutz Report 245 (1979)
[32] *E. A. Zherbin, A. B. Chukhlovin*, Tsitologiya **26** (1984) 973 (in Russian)
[33] *Yu. N. Anokhin, A. A. Yarilin*, Radiobiologiya **20** (1980) 391 (in Russian)
[34] *Yu. N. Anokhin, T. A. Norets, A. A. Yarilin, V. F. Stepanenko*, Radiobiologiya **23** (1983) 493 (in Russian)
[35] *S. J. Knox, M. Shifrine, L. S. Rosenblatt*, Radiation Research **89** (1982) 575
[36] *R. Kol, E. Ben-Hur*, Nicotinamide protects human lymphocytes from radiation inhibition of mitogenic stimulation. Proceedings of International Congress of Radiation Research, Amsterdam 1983, B6 – 18
[37] *G. Benga, R. P. Holmes*, Progress in Biophysics and Molecular Biology **43** (1984) 195
[38] *G. J. Köteles*, Acta Biochimica et Biophysica Acad. Sci. Hung., 1985 in press
[39] *M. Szamel, K. Resch*, J. of Biolog. Chem. **256** (1981) 11618
[40] *H. Wolters, D. Kelholt, A. W. T. Konings*, Radiation Research **102** (1985) 206
[41] *S. Suzuki*, Internat. J. Radiation Biology **47** (1985) 291
[42] *G. J. Schuurhuis, J. Hommes, J..Vos, I. Molenaar, A. W. T. Konings*, Internat. J. Radiation Biology **45** (1984) 159
[43] *R. A. Coss, J. R. Bamburg, W. C. Dewey*, Radiation Research **85** (1981) 99

[44] *S. J. Ladyman, K. M. S. Townsend, C. Edwards,* Radiation Research **99** (1984) 25

[45] *K. M. Yamada,* Annual Rev. Biochemistry **52** (1983) 761

[46] *M. Höök, L. Kjellén, S. Johansson,* Annual Rev. Biochemistry **53** (1984) 847

[47] *S. A. Newman, D. A. Frenz, J. J. Tomasek, D. D. Rabuzzi,* Science **228** (1985) 885

[48] *J. S. Rasey,* Cell-cell contact and repair of radiation damage. Proceedings of the International Congress of Radiation Research, Amsterdam 1983, B7 – R

[49] *D. P. Penney, W. A. Jr. Rosenkrans,* Radiation Research **99** (1984) 410

[50] ICRP Publication 41. Nonstochastic effects of ionizing radiation, Pergamon Press 1984

[51] *S. B. Field, A. C. Upton,* Internat. J. Radiation Biology **48** (1985) 81

[52] *R. Dover, C. S. Potten,* Internat. J. Radiation Biology **43** (1983) 681

[53] *J. O. Archambeau, G. W. Bennett, R. M. Shymko,* Radiation Research **94** (1983) 648

[54] *J. O. Archambeau, A. Ines, L. F. Fajardo,* Radiation Research **98** (1984) 37

[55] *D. F. Martin, J. J. Fischer,* Internat. J. Radiation Oncology, Biology and Physics **10** (1984) 1903

[56] *H. A. S. van den Brenk,* Internat. J. Radiation Biology **21** (1972) 607

[57] *H. S. Reinhold, G. H. Buisman,* British J. Radiology **46** (1973) 54

[58] *A. Dritschilo, T. Brennan, R. R. Weichselbaum, K. L. Mossman,* Radiation Research **100** (1984) 387

[59] *S. Mulgrew, M. Moriarty, A. Robinson,* Internat. J. Radiation Biology **46** (1984) 643

[60] *D. M. Peel, J. W. Hopewell, J. Wells, M. W. Charles,* Radiation Research **99** (1984) 372

[61] *D. M. Peel, J. W. Hopewell, J. Wells, M. W. Charles,* Radiation Research **101** (1985) 491

[62] *M. P. Law,* Advances in Radiation Biology **9** (1981) 37

[63] *G. S. Dimitrievich, K. Fischer-Dzoga, M. L. Griem,* Radiation Research **99** (1984) 511

[64] *H. Yamaura, T. Matsuzawa,* J. Radiation Research **25** (1984) 296

Isotopenpraxis 22 (1986) 10, S. 368 bis 371

Comparative Studies on the Development of a ^{188}W-^{188}Re Generator

N. Botros, M. El-Garhy, S. Abdulla, H. F. Aly[1])

(Nuclear Chemistry Department, Atomic Energy Establishment, Cairo)[1])

A comparative study on the sorption behaviour of tungsten and rhenium between anion exchange resin, charcoal and alumina and aqueous media of HNO_3, HCl and NaCl were carried out. The equilibrium distribution coefficient obtained for the different systems were evaluated and discussed in terms of the separation factor between W and Re. It is found that the systems based on alumina and HCl or NaCl are the most promising systems to be used for development of ^{188}W-^{188}Re generator. The capacity of a chromatographic column loaded with alumina for ^{188}W from 0·01 M HCl, 0·7% NaCl and 1·2% NaCl was found to equals 75·3, 91·0 and 126·0 mg/cm, respectively. A recommended procedure was adopted for ^{188}W-^{188}Re generator.

Es wurde eine Vergleichsuntersuchung zum Sorptionsverhalten von W und Re zwischen einem Anionenaustauscherharz einer Aktivkohle und einer Tonerde und den wäßrigen Lösungen von HNO_3, HCl und NaCl ausgeführt. Die Gleichgewichtsverteilungskoeffizienten der unterschiedlichen Systeme wurden bestimmt und als Trennfaktoren zwischen W und Re diskutiert. Es wurde gefunden, daß sich das System Tonerde/HCl oder NaCl sehr als Trennsystem für die Entwicklung eines ^{188}W-^{188}Re-Generators eignet. Die Kapazitäten einer mit Tonerde gefüllten Säule für ^{188}W mit 0·01 M HCl, 0·7% NaCl und 1·2% NaCl betrugen entsprechend 75·3, 90·0 und 126·0 mg/cm. Es wird eine geeignete Trennmethode für einen ^{188}W-^{188}Re-Generator empfohlen.

Keywords

aluminium oxides; charcoal; ion exchange; ion exchange chromatography; radioisotope generator; rhenium 188; separation processes; tungsten 188

Introduction

Radioisotope generators are of great importance and are widely used in medical and educational purposes [1, 2]. They overcome the transportation difficulties for direct supply of short lived radionuclides for a much longer periods than its own half life would permit. Another advantage of radioisotope generators is the possibility of using radionuclides at a location away from the place where it is produced. ^{188}W-^{188}Re is one of the examples where ^{188}Re is the daughter of ^{188}W via its β^- decay with half-life of 69 days. ^{188}W can be obtained by long exposure of W target to neutron irradiation [3]. The soft γ-rays emitted from ^{188}Re are mainly with 0·155 and 0·633 MeV and its relatively short half-life of 17 h, make it quite suitable for medical application in both radiodiagnostic and therapeutic purposes [1].

Some adsorption studies for W and Re were performed using anion exchanger and different aqueous media [4–9]. Other authors used the anion exchanger AV-17 in the EDTA form to study the behaviour of rhenium and other elements [10, 11]. Activated carbon either loaded with organic solvent [12] or platinated [13] was also used in separation of rhenium from different ions. Activated carbon adsorbs anions in air and the mechanism shows that rhenium is adsorbed in the form of ReO_4^- anion. ^{188}Re was separated from ^{188}W using ZrO_2 [3]. It was found that tungsten was strongly sorbed from weakly acidic and basic solutions. Rhenium was eluted by distilled water. Although rhenium was separated in a very pure state but its yield do not exceed 50%. The present work is directed to investigate the sorption behaviour of ^{188}Re and ^{188}W on anion

[1]) Address: Hot Laboratory Centre, P.O. Box 13759, Cairo, Egypt

exchanger, alumina for chromatography and activated charcoal from aqueous solutions of HNO_3, HCl and sodium chloride. The sorption results from different systems were compared in terms of their suitability for use as ${}^{188}W$-${}^{188}Re$ generator and the chosen system was chromatographically optimized to serve this purpose.

Experimental

γ-Counter: A γ-ray detector consists of a scintillation counter of the ECKO type and a NaI(Tl) 1 × 1 inch was used. Reagents: All the reagents were of high purity grade. They are B.D.H. or Merck products.

Radioactive ${}^{186-188}Re$ was obtained by irradiation of spectroscopic pure rhenium metal from *Johnson* and *Matthy* in the Et-RR-1 for 48 hours. It was then dissolved in 1 M HNO_3 in presence of few drops of H_2O_2. The final solution was 10^{-6} M obtained by dilution. ${}^{187}W$ was used to follow the behaviour of W, it was obtained by dissolving the irradiated sodium tungstate — after cooling time — in water to give a solution of 10^{-4} M.

Anion exchange resin (Ionenaustauscher III), alumina (900 active, 70—230 mesh, ASTM) and medical charcoal, from Merck, were used as the adsorbing materials. They were treated alternatively with HCl (0·1 M) NaOH (1 M) solution in order to free them from any impurities, finally with 0·1 M HCl solution and water till neutral. They were then dried, sieved and stored for use.

Batch Technique

A volume of 2 ml aqueous phase of certain concentration was shaked with 50 mg of the sorbent for about 2 h, which was found to be enough period to reach equilibrium. After the solid was settled down, 0·5 ml of the solution was pipetted and counted. The distribution coefficient was calculated from the equation:

$$K_d = \frac{A_0 - A}{A} \cdot \frac{v}{m}$$

where A_0 = initial activity in the aqueous phase,
A = activity of the solution after equilibrium,
v = the aqueous phase volume in ml and
m = is the solid phase weight in grams.

Column Technique

This technique was used to measure the alumina capacity for tungsten and elution study. A column of 2 cm height and 0·5 cm internal diameter with a tap was employed. It was filled with certain weight of alumina after incorporating glass wool in the bottom of the column. Before use, the column was conditioned with five column volumes of the eluent solution. The isotope to be studied was percolated through the column in a minimum volume of the solution and with constant flow rate of 0·2 ml/min. Subsequently known fraction volumes of the effluent were collected for radiometric assay.

Results and Discussion

Sorption studies for ReO_4^- and WO_4^{2-} were carried out using the sorbents, anion exchanger Ionenaustauscher III, active charcoal (100—250 μm) and alumina, Al_2O_3, 70 to 32 mesh and aqueous solutions of HNO_3, HCl and NaCl with different concentrations. The distribution coefficient data obtained together with separation factor values $\alpha^{WO_4^{2-}}_{ReO_4^-}$ which equals $K_{d\,WO_4^{2-}}/K_{d\,ReO_4^-}$ and represents the separation feasibility of the two radionuclides, are listed in Tab. 1—3. The variation of the K_d and $\alpha^{WO_4^{2-}}_{ReO_4^-}$ with HNO_3 concentration is given in Tab. 1. It is clear that in dilute acid concentration ≦ 0·01 the anion exchanger has a great affinity to adsorb WO_4^{2-} than ReO^- anion, the affinity begins to decrease at higher concentration from 0.1 M up to 1 M HNO_3 solutions and the $K_{d\,ReO_4^-}$ is higher than that of WO_4^{2-}. At higher acidities the K_d for both anions begins to decrease with the increase of acid molarities. The tendency of the anion exchanger used towards both anions under all acid concentrations used is relatively equals which is clear from the separation factor values α = 0·5 to 2·5. The distribution coefficient of WO_4^{2-} on charcoal is higher than the ReO_4^- distribution coefficient all over the HNO_3 acid concentration. Although the K_d values decrease with increase of the acid concentration yet the separation factor values goes higher with increasing acidity and $\alpha^{WO_4^{2-}}_{ReO_4^-}$ reaches its higher value at 8 M HNO_3.

In the case of alumina, nitric acid system, WO_4^{2-} is more strongly adsorbed than ReO_4^{2-} in dilute acid solution ($K_{d\,WO_4^{2-}}$ = 1000 while ReO_4^- = 20) giving $\alpha^{WO_4^{2-}}_{ReO_4^-}$ = 50 at 0.01 M acid solution. At concentrated acid solutions, the high value of the separation factor is due to the diminishing in K_d value for ReO_4^- (~ 0·01) and not due to the high $K_{d\,WO_4^{2-}}$ values.

Using HCl as the aqueous phase, the two anions behave in a similar way as in case of nitric acid medium. Charcoal tends to adsorb ReO_4^- in this medium than in case of HNO_3 while the K_d of WO_4^{2-} in HCl is less than in HNO_3, thus the separation factor is very small as given in Tab. 2, compared to that in Tab. 1 at the same concentration conditions. The selectivity of anion exchanger towards ReO_4^- reaches its maximum at 0·1 M HCl (low separation factor). At 8 M HCl the selectivity on the anion exchanger is reversed and high values of $\alpha^{WO_4^{2-}}_{ReO_4^-}$ is observed.

Alumina still prefer WO_4^{2-} than ReO_4^- but the separation factor is lower than in the case of HNO_3 aqueous phase specially at high HCl concentrations (from 2·0 M and more) where α do not exceed 4.4. Tab. 3 shows the variation in adsorbability of WO_4^{2-} and ReO_4^- towards the 3 sorbents from sodium chloride solution of concentrations range between 0·7—1·2%. The two anions were adsorbed on the anion exchanger with nearly equal K_d values allover the concentration range, thus the separation factor values are around one. The affinity of charcoal to adsorb ReO_4^- than WO_4^{2-} from NaCl solution is illustrated by $\alpha^{WO_4^-}_{ReO_4^-}$, value of order 10^{-3}. Contrary to charcoal, alumina selectivity for WO_4^{2-} from sodium chloride medium was more developed than in case of HNO_3 or HCl aqueous media. Thus very high separation factors were obtained, which equals 790 at 0·7% and reaches to 10 times that values when NaCl concentration increases to 1·2%.

Based on the aforementioned results specific systems can be suggested by choosing conditions of high separation factors. These systems are, 1- charcoal-HNO_3 system (6—8 M HNO_3), 2- anion exchange resin-HCl system (8 M HCl), 3- alumina HNO_3 system (0·1 M HNO_3), 4- alumina-HCl system (0·01 M HCl) and 5- alumina-NaCl

Tab. 1. Variation of K_d fo WO_4^{2-} and ReO_4^- with HNO_3 molarities and corresponding separation factor

Acid M	Anion exchanger			Active charcoal			Alumina		
	$K_{d\,WO_4^{2-}}$	$K_{d\,ReO_4^-}$	$\alpha^{WO_4^{2-}}_{ReO_4^-}$	$K_{d\,WO_4^{2-}}$	$K_{d\,ReO_4^-}$	$\alpha^{WO_4^{2-}}_{ReO_4^-}$	$K_{d\,WO_4^{2-}}$	$K_{d\,ReO_4^-}$	$\alpha^{WO_4^{2-}}_{ReO_4^-}$
0·01	3000	1500	2·0	5000	2500	2·0	1000	20·00	50
0·10	2000	4000	0·5	4500	1000	4·5	200	4·00	50
1·00	200	150	1·3	300	80	3·7	20	0·70	30
2·00	60	50	1·2	200	25	8·0	5	0·30	16
4·00	30	15	2·0	120	7·0	16·0	1·5	0·01	H.V.
6·00	16	7	2·3	150	2·5	60·0	0·7	0·01	H.V.
8·00	10	4	2·5	220	1·5	146·6	2·0	0·01	H.V.

Tab. 2. Variation of K_d of WO_4^{2-} and ReO_4^- with HCl molarities and corresponding separation factors

Acid M	Anion Exchanger			Active charcoal			Alumina		
	$K_{d\,WO_4^{2-}}$	$K_{d\,ReO_4^-}$	$\alpha^{WO_4^{2-}}_{ReO_4^-}$	$K_{d\,WO_4^{2-}}$	$K_{d\,ReO_4^-}$	$\alpha^{WO_4^{2-}}_{ReO_4^-}$	$K_{d\,WO_4^{2-}}$	$K_{d\,ReO_4^-}$	$\alpha^{WO_4^{2-}}_{ReO_4^-}$
0·01	40000	10000	4·000	300	3000	0·10	220·0	10·00	22·00
0·10	1500	80000	0·018	400	3500	0·10	90·0	8·00	11·25
1·00	250	1000	0·250	500	2500	0·20	20·0	1·50	13·30
2·00	170	500	0·340	550	1500	0·37	4·0	0·90	4·40
4·00	150	150	1·000	250	650	0·37	1·0	0·60	1·66
8 00	150	70	2·100	90	200	0·45	0·6	0·40	1·50
0·00	300	2	150·000	20	40	0·50	0·5	0·35	1·40

Tab. 3. Variation of K_d of WO_4^{2-} and ReO_4^- with sodium chloride concentration and corresponding separation factors

Sod. Cl. %	Anion exchanger			Active charcoal			Alumina		
	$K_{d\,WO_4^{2-}}$	$K_{d\,ReO_4^-}$	$\alpha^{WO_4^{2-}}_{ReO_4^-}$	$K_{d\,WO_4^{2-}}$	$K_{d\,ReO_4^-}$	$\alpha^{WO_4^{2-}}_{ReO_4^-}$	$K_{d\,WO_4^{2-}}$	$K_{d\,ReO_4^-}$	$\alpha^{WO_4^{2-}}_{ReO_4^-}$
0·7	5720	2414	2·40	2·5	2500	0·0010	1500	2·00	791·67
9·9	5624	4832	1·10	3·0	2400	0·0012	1800	3·50	513·61
1·0	4592	4998	0·91	2·8	2500	0·0011	2800	2·30	1235.46
1·2	4200	5115	0·82	1·9	2100	0·0009	10000	1·30	7692·31

system (allover the concentration range studied, specially 1·2% solution is preferred for its high separation factor.

Systems 1 and 2 were excluded as the possible separation due to the needs of high acid concentration which is not recommended for medical uses. Systems 3 and 4 were also refused even the separation can be done at low acid concentration, but the separation factors are rather small as $K_{d\,ReO_4^-}$ is rather high at such conditions which may complicate and affect Re-activity yield.

System 5 is the most suitable because of the high separation factor, and the low NaCl concentration eluent used as well as the very small $K_{d\,ReO_4^-}$ value $\leqq 3{\cdot}5$. Thus under these conditions, ^{188}Re is expected to be separated in small volumes of high chemical and radiochemical purity in a solution suitable for medical applications.

Adopted Procedure for Employing ^{188}W-^{188}Re Generator

From the previous results and discussion it was found that alumina is the suitable sorbent for W from dilute HCl or NaCl solutions. Thus, a column technique was developed to determine the break through capacity for W at eluent concentrations of high separation factors, (0·1 M HCl or 0·7 and 1·2% NaCl). The break through capacity values were calculated according to equation [14]:

$$Q = (C_0 \cdot V_{\text{free}})/L$$

C_0 is the initial concentration of tungsten in mg/cm³, V is the free column volume and L its length (in this experiment $V = 0{\cdot}6$ cm³ and $L = 2$ cm) and the elution curves are given in Fig. (1 and 2). The experimental conditions and corresponding Q values are given in Tab. 4, from

Tab. 4. Capacity values of the alumina column for tungsten eluted by hydrochloric acid or sodium chloride

Eluent conc., M or %	Initial conc. of W, mg/cm³	No. of free column volume	Capacity mg/cm
0·01 M HCl	35	2·15	75·2
0·7 % NaCl	35	2·60	91·0
1·2 % NaCl	35	3·60	126·0

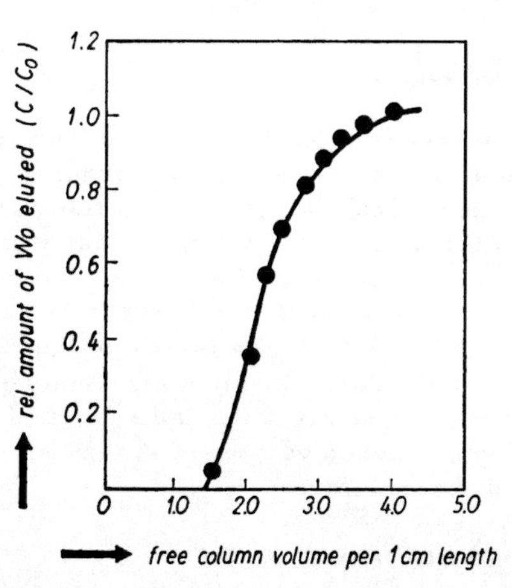

Fig. 1. Break through relation for the capacity of the alumina/HCl-system

Load solution $C_0 = 35$ mg/cm³
Eluent: 0·01 M HCl

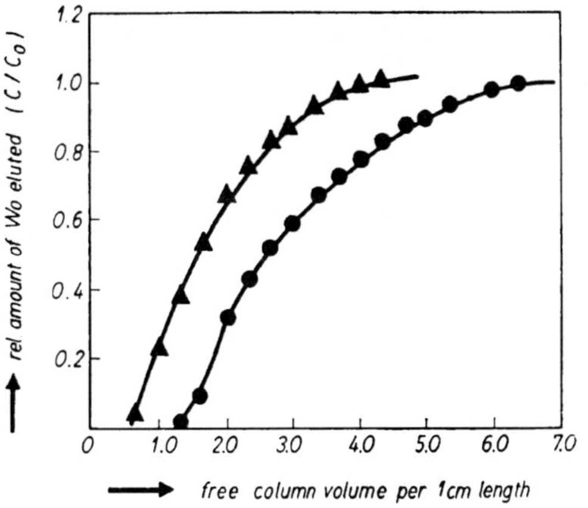

Fig. 2. Break through relation for the capacity of the capacity of the alumina/NaCl-system;

Load solution C_0 = 35 mg/cm³
–▲–▲– Eluent: 0·7 % NaCl; –●–●– Eluent: 1·2 % NaCl

which it is found that 1·2% NaCl solution is very suitable solution for loading W which is parallel to the batch technique results (high separation factor).

Based on these results a column of 10 cm height of alumina was used and loaded with W together with Re tracer in 30 cm³ of 1·2% NaCl solution. Re was found to be quantitatively eluted in the loading solution and no W was detected in the effluent. When further Re was added, no uptake of it was detected and ^{188}Re are completely passed through the column. From these results, the procedure adopted is summarized as follows: 1 g of sodium tungstate was irradiated for a long accumulative period of 30 days to obtain the parent ^{188}W according the reaction:

$$^{186}\mathrm{W}\ (\theta = 28{\cdot}4) \xrightarrow[\sigma=38]{n,\gamma} {}^{187}\mathrm{W}\ (T_{1/2} = 23{\cdot}8\ \mathrm{h})$$

$$\xrightarrow[\sigma=90]{n-\gamma} {}^{188}\mathrm{W}\ (T_{1/2} = 69\ \mathrm{d})$$

The irradiated target was then dissolved in 1·2% NaCl after suitable cooling time. This solution was fed to an alumina column of 10 cm height and 0·5 cm internal diameter. ^{188}W was completely adsorbed and ^{188}Re passed in the effluent. Repeated elution of ^{188}Re was almost quantitative with a yield of more than 98%.

Received December 7, 1985

References

[1] *M. Tubis, W. Wolf* (eds.), "Radiopharmacy", Wiley Interscience Publication, New York: J. Wiley & Sons 1976
[2] *J. Ingrand*, "Les Radioisotopes et leur Principles Applications Diagnostiques", Ch. Exploration de la Tyroide, Elements de Dosimetrie (in Press)
[3] *K. W. Malyshev, V. V. Smirnov*, Radiokhimiya 17 (1975) 249
[4] *H. Mori, Y. Fujimura, Y. Takegami*, Nippon Kagaku Kaishi 7 (1982) 1163
[5] *R. Kuroda, T. Isida, T. Seki, K. Oguma*, Chromatographia 15 (1982) 223
[6] *T. B. Shkodina, Kh. A. Al'Zhanova, V. G. Shkodin*, Deposited Doc., VINITI: 2997 (1982)
[7] *K. W. Cha, J. S. Choi*, Bull. Korean Chem. Soc. 2 (1981) 37
[8] *J. M. Peter, G. Del Fiore*, Radiochemical Radioanal. Lett. 21 (1975) 11
[9] *D. S. Gaibakyan, D. K. Khudaverdyan*, V. Sb. Renii, Khimiya, Tekhnol., Analiz. 66 (1976)
[10] *F. I. Brouchek, M. V. Mirianshivili, A. M. Saganelidze*, Soobshch. Akad. Nauk. Gruz. SSR. 81 (1976) 89
[11] *F. I. Brouchek, A. M. Saganelidze*, Izv. Akad. Nauk Gruz. SSR, Serkhim 1 (1975) 215
[12] *K. Kim Tai, J. E. Ritsko, M. B. Mac Innis, M. C. Vogt*, U.S.US. 4, 379, 126 (1983)
[13] *A. I. Loskutov, S. F. Andrianov, E. G. Siminov, I. A. Kuzin*, Zh. Prikl. Khim (Leningrad) 49 (1976) 729
[14] *R. G. Occult, W. J. Kaufman, G. Ksin*, AECU-3608, Report (1956)

Isotopenpraxis 22 (1986) 10, S. 371

Buchbesprechung

Atmosphärische Spurenstoffe und ihr physikalisch-chemisches Verhalten
Ein Beitrag zur Umweltforschung

Hrsg. K. H. Becker und J. Löbel. Berlin, Heidelberg, New York, Tokyo: Springer Verlag 1985. 264 S., 81 Abb., 27 Tab.; kart. DM 98.–

Die klassische Ausbildung von Naturwissenschaftlern und Chemikern behandelt die Inhaltsstoffe der Atmosphäre lediglich als Feststellung, daß sie vorhanden sind. Weder ihre Herkunft, noch ihr physikalisch-chemisches Verhalten werden eingehend erläutert; ihre Auswirkungen auf die belebte Natur sind nicht Gegenstand von Vorlesungen oder Seminaren. Dabei hat sich in den letzten Jahren ein Sonderzweig der Chemie herausgebildet, den man i. a. mit Luftchemie oder Aerochemie bezeichnet. Darüber gibt es bisher kein modernes einführendes Lehrbuch, um den Wissensbedarf abzudecken.

Insgesamt 12 Wissenschaftler haben sich vorgenommen, in 14 Abschnitten die wichtigsten Ergebnisse der Luftchemie geschlossen darzustellen. Neben dem kinetischen Verhalten der Spurengase, den ablaufenden Radikalreaktionen und chemischen Reaktionen in der unteren und oberen Atmosphäre werden die Aerosole, die Ablagerungen durch trockene Deposition und das Auftreten der Spurengase und Aerosole in den Niederschlägen behandelt. Einen breiten Raum nehmen Modellrechnungen über die Aufbereitung der Spurenstoffe ein, wobei die Vor- und Nachteile einzelner Ansätze beschrieben werden. Dem Charakter einer Einführung in die Luftchemie entsprechend, werden die Ableitungen der Grundgleichungen leicht überschaubar dargestellt und beispielsweise auch die unterschiedlich angewendeten Konzentrationsangaben übersichtlich zusammengefaßt.

Einen Beweis, wie schnell sich dieses Gebiet gerade in den letzten Jahren entwickelt hat, liefert auch die ausgewertete und weiterführende Literatur: von den insgesamt 245 angegebenen Literaturstellen entfallen allein auf die Jahre 1980–1984 96 Zitate (≙ 40%). Das im Vorwort formulierte Anliegen des Buches, die physikalisch-chemischen Zusammenhänge den interessierten Naturwissenschaftlern, Ingenieuren und Fachleuten, die mit Problemen des Umweltschutzes und der Reinhaltung der Luft befaßt sind, zu vermitteln, wird vollauf erfüllt, und es dürfte auch sehr geeignet sein, als Lehrbuch für Studierende und als Grundlage für Lehrende zu dienen.

H. Hübner

Isotopenpraxis 22 (1986) 10, S. 372 bis 375

Use of Radioactive Indicators for Investigation of Crystallization Processes in Founder's Alloys*)

L. H. Boyadjiev[1]), *Ts. Peneva*[1]), *H. Raichev*[2]), *L. Valish*[3]), *A. Kras*[3])

(Institute of Metal-Smelting and Founder Technique, Sofia, Bulgaria)[1])

(Institute for Nuclear Research and Nuclear Energy, Bulgarian Academy of Sciences)[2])

(Institute of Nuclear Chemistry and Engineering, Warsaw, Poland)[3])

The method of radioactive indicators and the method of autoradiography are the most convenient for determination of the crystallization front and the depth of the liquid phase. The aim of this work is to create an unified method for investigation of the processes of crystallization in continuous and semicontinuous founding of ferrous and non-ferrous metals. Au-198 was chosen as radioisotope. The techniques of investigation are described and some are discussed. The full picture of the form and the magnitude of the liquid phase in these technologies can be obtained only by means of this method.

Die Methode der radioaktiven Indikatoren und die Methode der Autoradiographie sind am geeignetsten, um eine Kristallisationsfront und die Tiefe der flüssigen Phase zu bestimmen. Das Ziel der vorgelegten Arbeit besteht in der Schaffung einer vereinheitlichten Methode zur Untersuchung von Kristallisationsprozessen bei einem kontinuierlichen und halbkontinuierlichen Schmelzen von Eisen- und Nicht-Eisen-Metallen. Als Radionuklid wurde Au-198 ausgewählt. Die Untersuchungstechnik wird beschrieben und die Ergebnisse an einigen Beispielen diskutiert. Das Gesamtbild der Form und das Ausmaß der flüssigen Phase bei diesen Herstellungsprozessen können nur mit Hilfe dieser Methode erhalten werden.

Keywords

autoradiography; brass; bronze; crystallization; gold 198; iron base alloys; nondestructive analysis; solidification; tracer techniques

1. Introduction

The continuous and semi-continuous founding is a method for obtainment of homogeneous structure of the metal alloys. These technologies influence considerably on the quality of the ready castings and lead to a decrease of the wastes in the mechanical treatment. It is important to know the form of the crystallization form and the depth of the liquid phase when acquiring these technologies as well as for the control of the production itself. This could give a possibility to define the influence of such technological parameters as: the way of feeding the liquid metal in a crystallizer, velocity of founding, temperature and debit of cooling water, as well as the region of secondary cooling in the process of hardening the metal, etc. on the quality of the castings.

The known methods for investigation the crystallization of metals casted in forms, as measurements the thickness of the solid crust in a given moment by pouring out of the rest liquid phase [1] and building of an isothermal curve at solidification by means of thermo-couples installed in the form [2], are inapplicable in the continuous and semi-continuous founding of the metals.

The literature data [3–5] and our preliminary experiments showed that the method of radioactive indicators and the method of autoradiography are the most convenient for determination of the crystallization front and the depth of the liquid phase.

In connection with the above mentioned and in the framework of the plan between the member countries in CMEA for 1981–1985 on application of the radioisotope methods in national economy, it was assigned to create an unified method for investigation of the processes of crystallization in continuous and semi-continuous founding of ferrous and non-ferrous metals.

2. Method of investigation

The radioactive indicator for autoradiographic investigation of the form and the size of the liquid metal phase in industrial conditions should be characterized with a comparatively small half-decay; to emit beta-particles; to have low radiotoxicity and good solubility in the alloy.

Gold-198 was chosen from the accessible radioisotopes which corresponds to the cited conditions and can be easily obtained with sufficient specific activity on (n, γ)-nuclear reaction.

Experimentally, the type of the autoradiographic material Agfa Gevaert D-7 structures was determined in laboratory conditions. This film is characterized with high sensitivity and allows to restrict to maximum the mass of the inserted radioactive indicator and to reduce almost 15 times the exposure time of the autoradiograms.

*) Presented at the XIVth International Symposium on "Autoradiography" November 19–23, 1984, Reinhardsbrunn, G.D.R.

[1]) Address: BG-Sofia, Boulevard Lenin 75, Bulgaria

2.1. *Technique of investigation in continuous founding of cast iron profiles*

The equipment for continuous founding of cast iron profiles (Fig. 1) consists of: metal-collector — 4, graphite crystallizer — 5 cooled with water where the surface of the solid crust of the profile — 6 is formed.

The radioactive indicator gold-198 with activity 0·25 to 0·3 Bq is melted with pure aluminium in ratio 1:3 and the obtained ligature is transferred into a brass ampoule. Immediately before use, the ampoule is installed on a steel rod — 2. The radioactive indicator is inserted into the mouth of the crystallizator in a preliminary chosen moment where it dissolves and distributes into the liquid phase.

After cooling, samples with thickness 0·02 m, longitudinally and transversally, are cut from the "labelled" part of the casting (about 1 m) as shown in Fig. 2.

The autoradiographic films are exposed on samples for 10–12 h.

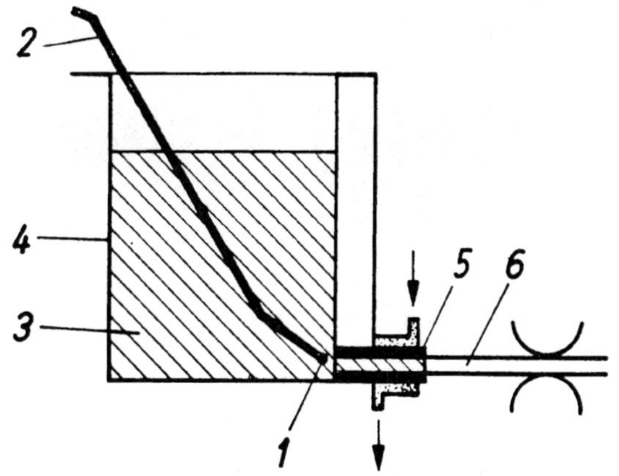

Fig. 1. Equipment scheme for continuous founding of cast iron profiles

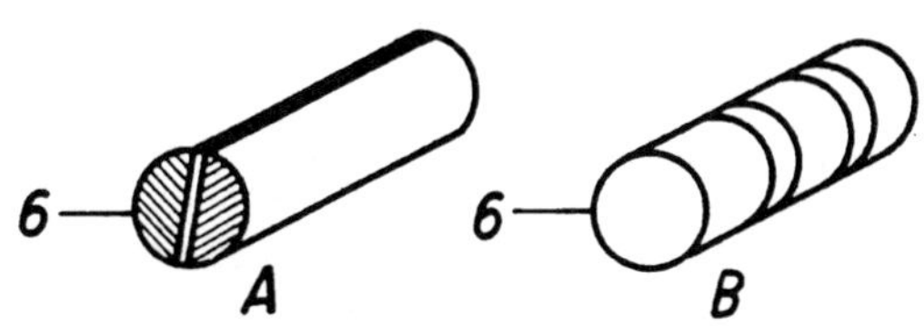

Fig. 2. Scheme for cutting samples from the cast iron profiles

2.2. *Investigation technique in semi-continuous founding of aluminium alloys*

The equipment for semi-continuous founding of aluminium slits with dimensions 0·3 × 1·1 up to 0·4 × 1·7 m consists of a region for feeding with liquid metal and a region for crystallization and secondary cooling.

The principle of passing the metal into the crystallizer is shown in Fig. 3. The radioactive indicator (Gold-198), preliminary dissolved in liquid aluminium, is introduced into the distributor. The activity of the radioactive indicator for "labelling" of separate regions of the slit is 3·7 Bq. Samples with thickness 0·02 m are cut by the scheme shown in Fig. 4 for autoradiographic investigations from the "labelled" region of the slit with length of about 0·6 m.

The autoradiograms are exposed on the samples for about 10 h.

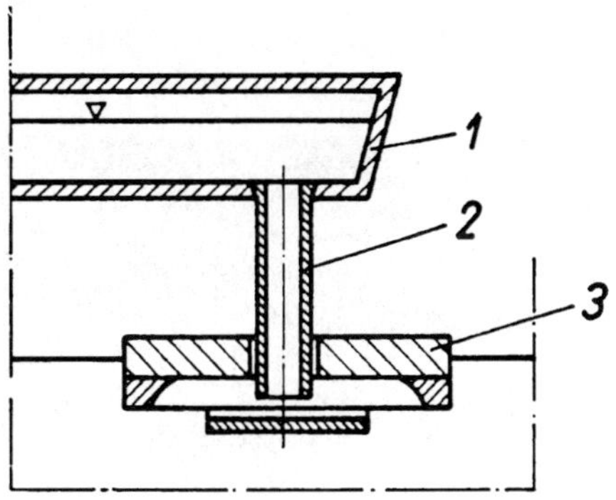

Fig. 3. Scheme for passing the liquid aluminium into the crystallizer

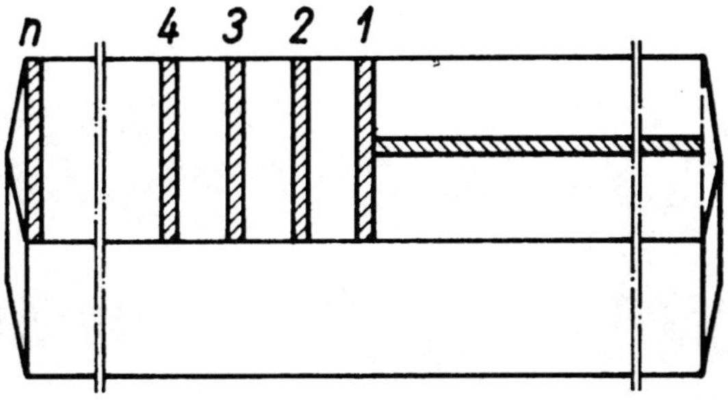

Fig. 4. Scheme for cutting samples with a thickness of 0.02 m

3. Experimental test of the method

The elaborated method for investigation of the crystallization front in continuous founding of cast iron and semi-continuous founding of aluminium was tested in industrial conditions in Poland and Bulgaria.

Slits of cast iron — 1 of diameter 0·095 m were chosen for experimental test of the method in continuous founding of iron cast, and in the semi-continuous founding of aluminium (99·5% Al) — slits with dimensions 0·40 × 1·7 m.

The autoradiograms of the "labelled" liquid phase for three arbitrarily chosen cases in continuous founding of cast iron profiles are presented in Fig. 5–7 of equal scale. The autoradiograms in Fig. 8–10 show the limit between the liquid and the solid phases of aluminium — cross section at distances 0·350, 0·720 and 0·09 from the external side of the slit.

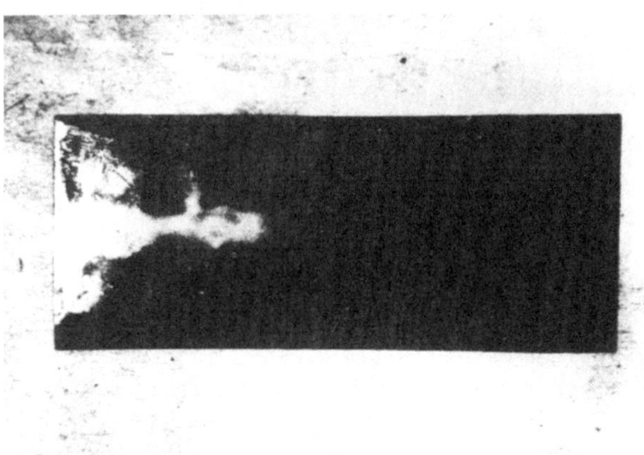

Fig. 5. Autoradiogram — longitudinal section of cast iron profiles ∅ 0·095 m, moderation time 6 s

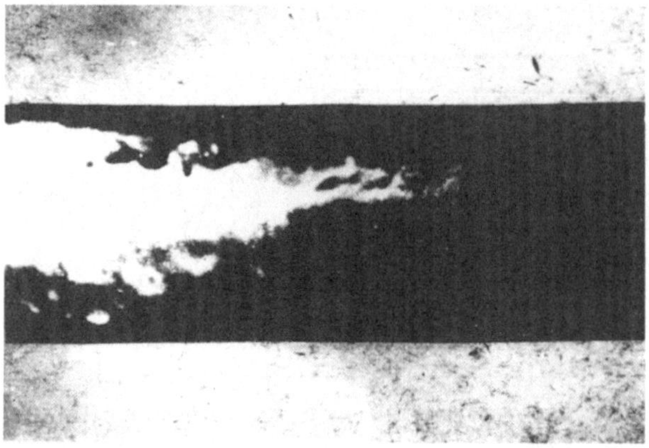

Fig. 6. Autoradiogram – longitudinal section of a cast iron profile ∅ 0·095 m, moderation time 9 s

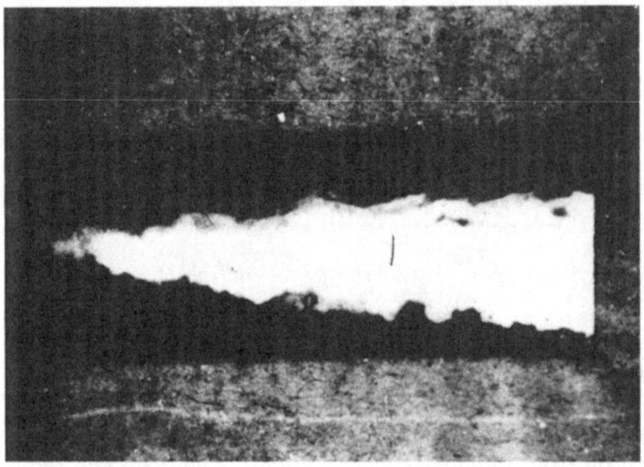

Fig. 7. Autoradiogram – longitudinal section of a cast iron profile ∅ 0·095 m, moderation time 12 s

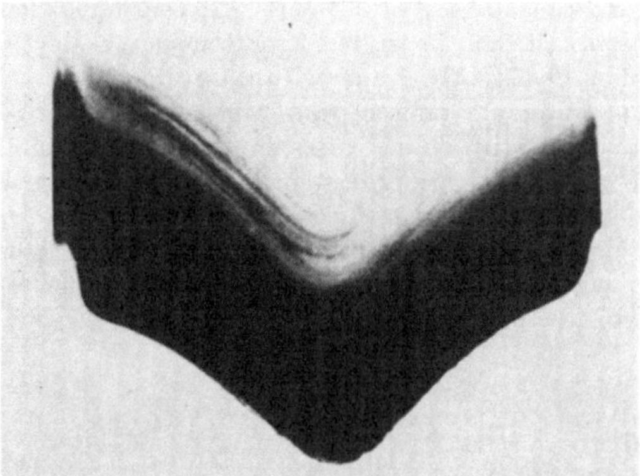

Fig. 8. Autoradiogram – longitudinal section of aluminium slit 0·4 × 1·7 m, distance from the wall 0·85 m

In the autoradiograms (Fig. 5–10), regions are seen with different optical density of blackening (the image in Fig. 5–7 is negative). This property to blacken allows to get information about the crystallization processes in solidification of the metal slits.

The external contour of blackening in the autoradiograms corresponding to the maximum accumulation of the indicator, is accepted to be the limit between the solid and the liquid phase for determination of the crystallization front and the depth of the liquid phase.

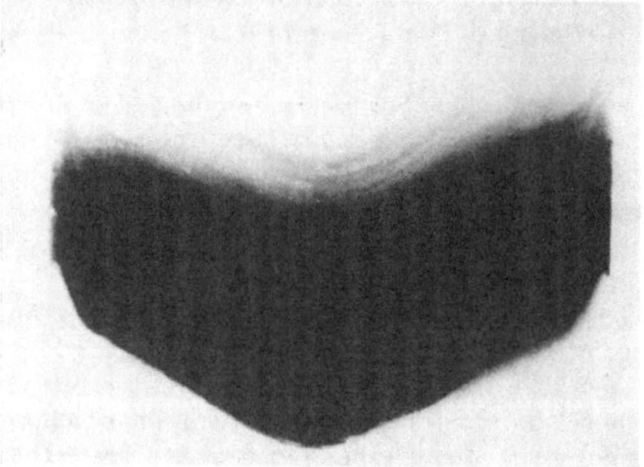

Fig. 9. Autoradiogram – longitudinal section of aluminium slit 0·4 × 1·7 m, distance from the wall 0·72 m

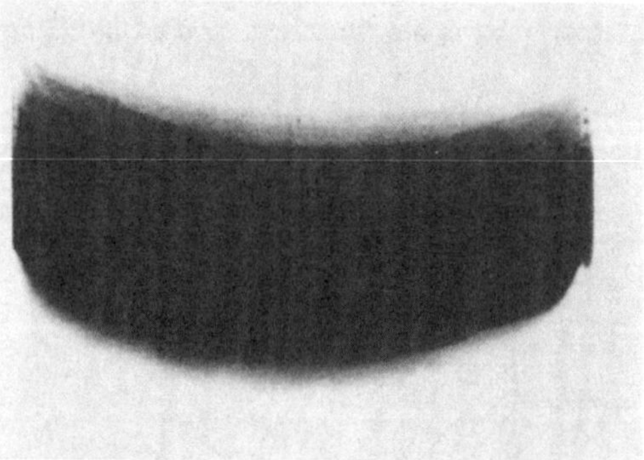

Fig. 10. Autoradiogram – longitudinal section of aluminium slit 0·4 × 1·7 m, distance from the wall 0·09 m

The local velocities of metal solidification can also be defined by the profile of the crystallization front, and together with the results from the metalographic investigations give a possibility to estimate the technology of founding.

4. Discussion

The autoradiograms in Fig. 5–7 show the limit between the solid and liquid phases of the cast iron profiles obtained at three different moderation times – 6, 9 and 12 s and at the same time of drawing out – 8 s.

At longer moderation, the depth of the liquid phase is smallest and at shorter time of moderation the liquid phase has the biggest depth.

The direct consideration of the radiograms gives an opportunity to draw also other interesting conclusions, for instance about the action of the crystallizer. In some cases, the uneven cooling can be the reason for worsening the homogeneity of the founded structure. Also particles are observed of freshly formed crust carried along under the influence of the metal static pressure of the liquid metal.

The analysis of the obtained autoradiograms in semicontinuous founding of aluminium helps to get a spatial picture of the crystallization front and to make an estimation for the influence of the different technological parameters.

It is established that for slits with equal dimensions and velocity of founding, the way of passing the liquid metal influences greatly the crystallization front. The use of two floats increases the depth of the liquid phase in the central part of the slit in comparison to the pass of the liquid metal through one of the floats.

It is also found that the form of the crystallization front close to the side walls of the slit depends on the velocity of founding.

It follows from the obtained results that the influence of the technological factors on the process of crystallization could not be considered separately. The changes of one of them leads in all cases to a correction of the rest as well.

Conclusion

An unified method is worked out for investigation of the crystallization processes in semi-continuous founding of aluminium slits with big dimensions and in continuous founding of cast iron profiles by radioactive indicators and autoradiograms.

The full picture of the form and the magnitude of the liquid phase in these technologies can be obtained only by means of this method.

Received October 17, 1985

References

[1] *V. V. Shepelev* et al., Problems of steel slit, Moscow 1974, 268–271
[2] *V. I. Dobatkin* et al., Founding production **10** (1969)
[3] *J. L. Putman*, J. Inst. Metals **81** (1954) 411
[4] *E. Budjaso*, Femipar Kutato Int. Kösi **3** (1969) 108
[5] *K. Akerman* et al., Isotopenpraxis **3** (1967) 211

Isotopenpraxis 22 (1986) 10, S. 375

Kurze Originalmitteilung

Synthese radioaktiv markierter Verbindungen. 54. Mitt.*) Darstellung von [3,5-^{14}C]3,5-Dibrom-4-hydroxybenzonitril

B. Schneider, H.-R. Schütte, H.-J. Schöbel[2])

(Akademie der Wissenschaften der DDR, Institut für Biochemie der Pflanzen, Halle/S.)[1])

(VEB Synthesewerk Schwarzheide, Kombinat SYS)[2])

Keywords

carbon 14; labelled compounds; labelling; organic compounds; synthesis

Für pflanzenphysiologische Untersuchungen benötigen wir [^{14}C]-markiertes 3,5-Dibrom-4-hydroxybenzonitril. Da [2,6-^{14}C]4-Aminophenol zur Verfügung stand, erschien die Umsetzung in einer Sandmeyerreaktion zum [3,5-^{14}C]4-Hydroxybenzonitril und anschließende Bromierung zum [3,5-^{14}C]3,5-Dibrom-4-hydroxybenzonitril als die günstigste Synthesevariante [2].

Zu einer Lösung von 500 mg $CuSO_4 \cdot 5\,H_2O$ (2 mmol) in 2,4 ml Wasser werden unter Rühren bei Raumtemperatur 420 mg (8,6 mmol) NaCN zugegeben. Nach dem Erhitzen auf 90 °C wird die aus einer vorgewärmten Lösung von 218 mg (2 mmol) [2,6-^{14}C]4-Aminophenol (130 MBq pro mmol) in 1,6 ml Wasser und 355 µl konzentrierter HCl sowie 138 mg (2 mmol) $NaNO_2$ in 0,4 ml Wasser bereitete Diazoniumchloridlösung zugefügt und 15 min gerührt. Nach dem Abkühlen werden unter langsamem Rühren 0,5 ml konzentrierte HCl zugegeben, wobei sich das Reaktionsgemisch aufhellt. Die Lösung wird durch Zugabe von festem NaCl gesätigt. und anschließend zehnmal mit je 6 ml Ether extrahiert. Der Etherextrakt wird filtriert, mit Molsieb (Wolfen-Zeosorb 3A, VEB Laborchemie Apolda) getrocknet und nach Einengen über Kieselgel (Kieselgel 60 reinst, 0,063–0,2 mm, Fa. Merck, Säule 25 × 3 cm, Elution mit Benzen:Aceton 3:1) chromatographiert. Die radioaktiven Fraktionen werden durch Sublimation im Vakuum gereinigt (4 hPa/120 °C). Dünnschichtchromatographie (Silufol UV_{254}, Fa. Kavalier, ČSSR, Laufmittel 1: n-Hexan:Aceton 3:1, 3 × entwickelt, R_f 0,23 und Laufmittel 2: Chloroform:Ether:Eisessig 40:9:1, R_f 0,60) zeigte die radiochemische Reinheit des erhaltenen [3,5-^{14}C]4-Hydroxybenzonitril.

Spezifische Radioaktivität 127 MBq/mmol

Ausbeute: 205 mg (86%)

Zur Bromierung wird der Lösung des [3,5-^{14}C]4-Hydroxybenzonitril in 2 ml 1-Propanol unter Kühlen und Rühren 200 µl Brom zugetropft. Anschließend wird bei 40 °C 3 h gerührt, wobei das Produkt teilweise kristallisiert. Das Reaktionsgemisch wird in 60 ml Wasser gegeben, das 0,5 ml Natriumhydrogensulfitlösung enthält, und nach Absaugen zweimal mit je 5 ml Wasser gewaschen. Zur Reinigung wird im Vakuum sublimiert (2,67 hPa/135 °C) und aus 5 ml Methanol/Wasser 3:2 umkristallisiert.

Die radiochemische Reinheit des erhaltenen [3,5-^{14}C]-3,5-Dibrom-4-hydroxybenzonitril beträgt > 99% und wurde unter den oben beschriebenen Bedingungen ermittelt (Laufmittel 1: R_f 0,34, Laufmittel 2: R_f 0,70).

Spezifische Radioaktivität: 127 MBq/mmol

Ausbeute: 383 mg (69% bezogen auf p-Aminophenol)

Unter gleichen Bedingungen und mit gleichem Ergebnis gelang die Synthese von [3,5-^{14}C]3,5-Dibrom-4-hydroxybenzonitril der spezifischen Radioaktivität 1,24 GBq pro mmol.

*) 53. Mitt. vgl. [1]

[1]) Anschrift: DDR-4050 Halle/S., Weinberg 3

Literatur

[1] *A. Unverricht, D. Groß*, J. labelled Comp. and Radiopharmaceuticals im Druck.
[2] *J. L. Buckland, R. F. Collins, E. M. Pullin*, Pestic. Sci. **4** (1973) 149.

Isotopenpraxis 22 (1986) 10, S. 376

Buchbesprechungen

Radiopharmaceuticals and Labelled Compounds 1984

Vienna: IAEA 1985. 581 S., 203 Abb. und diverse Tab., ÖS 1.140,–

Der Band stellt die Proceedings der „International Conference on Radiopharmaceuticals and Labelled Compounds" dar, die vom 22.–26. Oktober 1984 in Tokio stattfand und von der Internationalen Atomenergiekommission organisiert worden war. Das Anliegen der Konferenz war es, die jüngsten Entwicklungen auf diesem Gebiet vorzustellen, die sich seit dem letzten Symposium ergeben haben, das 1973 in Kopenhagen stattfand. Die Radiopharmazie hat sich seitdem dramatisch entwickelt, und neue wissenschaftliche Konzeptionen und Innovationen wurden eingeführt. Damit hat die diagnostische Nuklearmedizin einen ungeheuren Fortschritt erlebt. Die vorliegenden Proceedings behandeln Untersuchungen über die Beziehung zwischen chemischer Struktur und biologischer Aktivität von markierten Verbindungen sowie Experimente, die zu einer verbesserten Erkenntnis der Chemie und Biochemie von Radiopharmazeutika führen. Die verschiedenen Beiträge sind thematisch in neun Gruppen zusammengefaßt. In der ersten Gruppe wurden fünf Vorträge zur Produktion von Radionucliden für medizinische Anwendung gehalten. Die zweite Gruppe behandelt mit vier Beiträgen Radionuclidgeneratorsysteme. Die dritte Gruppe umfaßt drei Beträge zur Markierung und Anwendung von monoclonalen Antikörpern. Viele Beiträge beschäftigen sich mit der Chemie und Anwendung von Technetium-99m-Radiopharmazeutika, was die Bedeutung dieses Radionuclids für die Routine nuclearmedizinischer Diagnostik unterstreicht. So sind in Gruppe 4 sieben Beiträge zur Chemie von Technetium-99m-Radiopharmazeutika und in Gruppe 5 fünf Beiträge zur Anwendung neuer Technetium-99m-Radiopharmazeutika zusammengefaßt. Die Synthese von markierten Verbindungen ist mit sechs Beiträgen in Gruppe 6 vertreten und drei Beiträge behandeln in Gruppe 7 die Herstellung von Tracern und die Qualitätskontrolle für Radioimmunoassays. Die letzten beiden Gruppen betreffen mit drei Arbeiten das Thema „Der gegenwärtige Stand und zukünftige Aspekte der Radiopharmazeutika" sowie mit elf Beiträgen das Thema „Verteilungsuntersuchungen mit Radiopharmazeutika".

Ein breiter Raum wird auch von den vielversprechenden Anwendungen solch kurzlebiger Elemente wie ^{11}C, ^{13}N, ^{15}O sowie ^{18}F eingenommen. Schließlich sind 20 Posterbeiträge zu den genannten Themen enthalten. Die meisten Beiträge sind in englischer Sprache verfaßt. Jede Arbeit enthält ein ausführliches Literaturverzeichnis. Letztlich enthält der Band die Teilnehmer des Symposiums mit ihren Adressen. Die ausgewogene Zusammenstellung von zusammenfassenden Überblicksartikeln und Experimentalarbeiten macht den Band zu einem wichtigen Werkzeug für alle, die sich mit Synthese und Anwendung von Radiopharmazeutika und verwandten Substanzen beschäftigen. Er wird damit all solchen Interessenten als Bereicherung empfohlen.

H.-R. Schütte

Nukleare Festkörperphysik

G. Schatz, A. Weidinger. Stuttgart: B. G. Teubner 1985. 278 S., 158 Abb., 9 Tab. (Teubner Studienbücher Physik); kart. DM 29,80

Die Aufgaben der zwischen Kernphysik und Festkörperphysik angesiedelten nuklearen Festkörperphysik bestehen in erster Linie darin, Festkörpereigenschaften mittels verschiedener kernphysikalischer Methoden zu untersuchen. Nukleare Meßverfahren zeichnen sich in der Regel durch extreme Empfindlichkeit und hohe Meßgenauigkeit aus. Aus diesem Grund haben sie insbesondere in der modernen Festkörperanalytik eine große Bedeutung erlangt.

Im deutschsprachigen Raum fehlte bisher eine lehrbuchhafte Darstellung der nuklearen Festkörperphysik. Diese Lücke ist mit dem vorliegenden Buch nunmehr geschlossen. Es umfaßt zehn Kapitel. Auf eine kurze Beschreibung der elektrischen und magnetischen Kernmomente, der elektromagnetischen Kernübergänge sowie der Messung von Gammastrahlung folgt die Behandlung der Hyperfeinwechselwirkung. Dieses Phänomen bildet für die meisten der dargestellten Methoden die Grundlage.

Die Hauptkapitel des Buches sind folgenden Möglichkeiten der nuklearen Festkörperuntersuchung gewidmet: Mößbauer-Effekt, gestörte γ-γ-Winkelkorrelation (PAC), magnetische Kernresonanz (NMR), Kernorientierung (NO), Myon-Spin-Rotation (μSR), Positronenvernichtung und Neutronenstreuung. Neben den theoretischen Grundlagen und den meßtechnischen Voraussetzungen jeder Methode werden typische Anwendungsbeispiele aus der Festkörperphysik diskutiert. Es ist bedauerlich, daß der Channeling-Effekt und die Rutherford-Rückstreuung keine Aufnahme finden konnten.

Das Buch ist vor allem für Physikstudenten höherer Semester bestimmt. Es setzt solide Kenntnisse der wesentlichen Grundtatsachen der Kern- und Festkörperphysik voraus. Durch den klaren und systematischen Aufbau, die exakte Darstellung sowie zahlreiche aussagekräftige Abbildungen vermittelt es einen sehr guten, wenn auch knappen Überblick über die angedeuteten Probleme.

Als Einführung in das Gebiet der nuklearen Festkörperphysik ist das Buch ausgezeichnet geeignet.

W. Stolz

ISOTOPENPRAXIS

Band 22 1986 Heft 10

Contents / Inhalt

►Language of Publication ►Publikationssprache

Isotopenpraxis is indexed in Current Contents/PC & ES, Science Citation Index, and INIS Atomindex